Progress in Molecular and Subcellular Biology 16

Series Editors

Ph. Jeanteur, Y. Kuchino,
W.E.G. Müller (*Managing Editor*)
P.L. Paine

Springer
Berlin
Heidelberg
New York
Barcelona
Budapest
Hong Kong
London
Milan
Paris
Santa Clara
Singapore
Tokyo

Y. Kuchino   W.E.G. Müller   (Eds.)

Apoptosis

With 36 Figures

Springer

Prof. Dr. Y. KUCHINO
National Cancer Center Research Institute
Dept. of Biophysics
Tsukiji 5-Chome
Chuoko, Tokyo
Japan

Prof. Dr. W.E.G. MÜLLER
Universität Mainz
Institut für Physiologische Chemie
Abteilung für Angewandte Molekularbiologie
Duesbergweg 6
55099 Mainz
Germany

ISBN 3-540-59470-1 Springer-Verlag Berlin Heidelberg New York

Library of Congress Cataloging-in-Publication Data. Apotosis/Y. Kuchino, W.E.G. Müller, (eds.) p. cm. – (Progress in molecular and subcellular biology: 16) Includes bibliographical references. ISBN 3-540-59470-1 (hc) 1. Apoptosis. 2. Pathology, Cellular. I. Kuchino, Y. (Yoshiyuki), 1941– . II. Müller, W.E.G. (Werner E.G.). 1942– . III. Series. QH671. A652 1996 574.87'65 – dc20 95-37396

Springer-Verlag Berlin Heidelberg 1996
Printed in Germany

Cover design: Springer-Verlag, Design & Production

Typesetting: Thomson Press (India) Ltd., Madras

SPIN: 10135522   39/3132/SPS – 5 4 3 2 1 0 – Printed on acid-free paper

Preface

Apoptosis is an active form of cell death in most multicellular organisms, and plays an important role in different biological processes, such as remodeling of tissues in embryonic development, clonal selection of specifically reactive population of lymphocytes, elimination of damaged cells during hematopoiesis, and normal tissue turnover. Apoptosis is also observed in tumor cells treated with a variety of cancer therapeutic agents such as cytotoxic drugs, hormones, and irradiation. Thus, apoptosis has aroused the interest of a variety of scientists, including clinicians, and has become a fashionable subject of general biological studies. In this book, new information on the mechanism and regulation of apoptosis is summarized. It is hoped that the book will be useful to all working in the area of apoptosis.

Apoptotic cell death induced in various tissues and cells through multiple signal transduction pathways appears to be a cellular event deviating from tissue homeostasis, which depends on a balance between cell proliferation and death. It has been known that the balance of homeostasis is maintained by expression regulation of cellular genes including oncogenes and oncosuppressor genes. Indeed, expression of genes such as *c-myc*, *bcl-2* and *p53* influences cellular susceptibility to undergo apoptosis. For instance, depending upon the availability of critical growth factors, *c-myc* expression seems to determine the fate of the cell, that is continuous proliferation or apoptosis. By addition of serum growth factors, immortalized fibroblasts such as Rat-1a and NIH3T3 cells maintain the proliferation cycle. However, withdrawal of serum growth factors, which is not accompanied by growth arrest, induces apoptosis in the cells modified to express c-Myc constitutively. Expression of wild-type p53 also induces apoptosis in susceptible cells. Recent works suggest that temporary growth arrest induced by wild-type p53 cooperates with endogenous c-Myc in initiation of apoptosis. In this regard, it is of particular interest that constitutive *s-myc* expression, which causes arrest in G1 phase of cell cycle in glioma cells, can induce apoptosis in the tumor cells even in the presence of serum growth factors.

Both Myc and p53 function as transcription factors. Expression of these genes, which may be activated by stimuli given to cells, could induce new transcription of genes which may be triggered to undergo apoptosis. Such apoptosis induction, requiring de novo synthesis of new gene products, is generally blocked by metabolic inhibitors such as actinomycin D or cycloheximide. However, under certain circumstances, actinomycin D and cycloheximide do not inhibit apoptosis. The monoclonal antibody anti-Fas induces apoptosis in some cell types, causing cell death within 5–6 h. The apoptotic signal through Fas is induced by the binding of the Fas ligand, which is a TNF-related molecule of 40 kDa to Fas. Interestingly, the Fas-mediated apoptosis in some cells, such as mouse primary hepatocytes and mouse fibroblast cells, is enhanced rather than inhibited by the metabolic inhibitors. The Fas ligand is expressed in some CTL (cytotoxic T-lymphocyte) cell lines and in activated splenocytes. Therefore, it is believed that the Fas system should have an important role in CTL-mediated cytotoxicity.

Apoptosis in lymphocytes is induced not only by CTL-target cell interaction, irradiation, chemotherapy, hormone treatment, and oxidative or thermal stress, but also by infection with retroviruses like HIV and FeLV. The retrovirus infection modulates $CD4^+$ and $CD8^+$ T cell functions and leads to an acute decrease in the number of these T-cells by apoptosis induction. These studies on the retrovirus-mediated apoptosis may lead to a further understanding of the mechanism of the cytopathic effects of retroviruses.

Some cellular and viral genes appear to have an important role in protecting cells against apoptosis. The *bcl-2* gene is a typical gene whose expression blocks apoptosis in some cell types. It was isolated from the breakpoint of the translocation between chromosomes 14 and 18 found in human lymphomas. Initially, it was found that *bcl-2* expression effectively rescues myeloid precursors and pre-B cells undergoing apoptotic cell death upon growth factor withdrawal. Further gene transfection studies showed that *bcl-2* expression can inhibit apoptosis induced by several different types of stimuli. Recently, using transgenic mice, researchers showed that the topic expression of the *bcl-2* gene can result in a condition mimicking an autoimmune disease. In contrast, using *bcl-2*-deficient mice, they indicated that *bcl-2* deficiency does not interfere with embryonic development, but that abnormalities in kidney and other organs become evident soon after birth.

Although there are multiple ways to initiate the apoptotic pathway, in the final stage of the apoptotic pathway, several common features such as progressive fragmentation of chromosomal DNAs and morphological changes are observed in cells. For several years, the endonuclease involved in apoptosis has been considered to be Ca^{2+}/Mg^{2+}-dependent, its activity being increased by intracellular Ca^{2+}. However, recent studies show that there are several different types of endonucleases involved in the DNA fragmentation in apoptotic cells: Ca^{2+}/Mg^{2+}-dependent endonuclease; DNase II, Ca^{2+}-independent acidic endonuclease activated by decreasing intracellular pH and DNase γ. These findings suggest that activation of an endonuclease in apoptotic cells is closely associated with the extracellular environment.

As described above, apoptosis plays a central role in the regulation of both normal and malignant cell proliferation. Disrupted regulation of this control mechanism may cause serious human diseases such as encephalomyelitis and cancer. Therefore, understanding of the molecular mechanisms of apoptotic cell death and identification of the cell-type-specific factors determining cellular survival or death should lead to fundamental advances in the therapy of human diseases, including cancers.

Tokyo and Mainz, March 1995   Y. KUCHINO
                                                           W.E.G. MÜLLER

Contents

An Endonuclease Responsible for Apoptosis 1
S. Tanuma and D. Shiokawa

1   Introduction 1
2   Results 2
3   Discussion 6
References 10

**Cytopathic Feline Leukemia Viruses Cause Apoptosis
in Hemolymphatic Cells** 13
J.L. Rojko, J.R. Hartke, C.M. Cheney, A.J. Phipps,
and J.C. Neil

Abstract 13
1   Introduction 13
2   Cytopathic Infections with FeLV-C in Vivo
      and Direct Induction of Apoptosis in T4 Cells
      by FeLV-C 16
3   Killing of Lymphocytes by FeLV-FAIDS/p61C
      and Associated Variants 17
4   FeLV-C-Related Nonregenerative Anemia:
      Potential Role for Apoptosis 22
5   FeLV Genes Important in Cytopathicity/Apoptosis
      Induction 24
5.1   General Remarks 24
5.2   T-Cell Killing 24
5.3   Erythroid Aplasias 29
6   Speculative Relationship to Endogenous Virus
      Recombinants and Importance Across Evolution 30
7   Regions of FeLV p15E Important
      in Immunosuppression 31
8   Other Types of Cytopathic Disease Seen in Viremic
      Cats Which Could Have an Apoptotic Basis 31
8.1   Enteritis 31
8.2   Infertility and Abortion 32
8.3   Neurologic Syndrome 32
9   So How Does FeLV Cause Apoptosis? 33

10   Other Retroviruses and Animal Viruses
          as Potential Causes of Apoptosis   33
10.1   Retroviruses that Behave as Superantigens   33
10.2   HIV and Other Lentiviruses   34
10.3   Chicken Anemia Virus.   35
References   35

Neurotoxicity in Rat Cortical Cells
Caused by N-Methyl-D-Aspartate (NMDA) and gp120
of HIV-1: Induction and Pharmacological Intervention   44
W.E.G. MÜLLER, G. PERGANDE, H. USHIJIMA, C. SCHLEGER,
M. KELVE, and S. PEROVIC

Abstract   44
1   Introduction   45
2   Induction of Apoptosis in Rat Cortical Cells
          by HIV-1 gp120 or NMDA In Vitro   46
2.1   DNA Fragmentation and Cell Morphology
          After gp120 Treatment   46
2.2   Induction of Apoptosis in Cortical Cell Cultures
          by HIV-1 Particles   46
2.3   Increased Release of Arachidonic Acid
          from Neurons After Incubation with gp120.   48
2.4   Inhibition of Arachidonic Acid Release
          by Phospholipase A$_2$ Inhibitor   48
2.5   Induction of Apoptosis in Cortical Cell Cultures
          of NMDA.   49
2.6   Influence of NMDA Antagonists on Arachidonic
          Acid Release.   49
2.7   Arachidonic Acid Augments the NMDA-Caused
          DNA Fragmentation   51
3   Prevention of Apoptosis in Cortical Cells In Vitro
          by Memantine and Flupirtine   51
3.1   Memantine   51
3.2   Flupirtine   52
3.2.1   Prevention of NMDA or HIV-gp120-Induced
          Apoptosis in Cortical Cells by Flupirtine   52
3.2.2   Cytoprotective Effect of Flupirtine
          on Untreated Rat Cortical Cells   52
4   Conclusion   53
4.1   Cell Biological Findings   53
4.2   Pharmacological Interventions   54
References   55

**Apoptosis of Mature T Lymphocytes: Putative Role
in the Regulation of Cellular Immune Responses
and in the Pathogenesis of HIV Infection** 58
D. KABELITZ, T. POHL, H-H OBERG, K. PECHHOLD,
T. DOBMEYER, and R. ROSSOL

Abstract 58
1   Apoptosis of Immature T Lymphocytes 58
2   Apoptosis of Mature T Lymphocytes 59
3   Apoptosis of T Lymphocytes Induced
       by Superantigens and Conventional Antigen 60
4   Apoptosis of Mature T Lymphocytes
       In Vivo and Ex Vivo 62
5   Role of T-Cell Apoptosis in HIV Pathogenesis 63
6   Concluding Remarks 66
References 67

***bcl-2*: Antidote for Cell Death** 72
Y. TSUJIMOTO

Abstract 72
1   Introduction 72
2   Structure and Expression of the *bcl-2* Gene 73
3   Biological Function of *bcl-2*: Apoptotic
       Death-Sparing Activity 74
4   Role of *bcl-2* in the Immune System 74
4.1   *bcl-2* in Lymphocyte Selection 74
4.2   Production of Autoimmune Disease 75
5   Role of *bcl-2* in Neuronal Tissue 75
6   Role of *bcl-2* in Lymphomagenesis 76
7   Role of *bcl-2* in Other Systems 77
7.1   Virus Persistent Infection 77
7.2   Epithelial Cells 78
7.3   Morphogenesis 78
8   Analysis of *bcl-2*-Deficient Mice 78
9   Subcellular Localization of *bcl-2*: Multiple Membrane
       Locations 79
10   Biochemical Function of bcl-2 Protein 80
11   *bcl-2* Related Genes and bcl-2 Associated Proteins ... 81
12   Epilogue 82
References 82

Apoptosis Mediated by the Fas System 87
S. NAGATA

Abstract 87
1   Introduction 87
2   Fas, a Receptor for a Death Factor 88

2.1   Molecular Properties of Fas 88
2.2   Expression of Fas 90
3   Mutation in the Fas Gene of lpr-Mice 90
3.1   Chromosomal Gene for Fas 90
3.2   Insertion of an Early Transposable Element
        in Intron of Fas Gene in *lpr*-Mice 90
3.3   A Point Mutation in the Fas Gene of *lpr^{cg}* Mice 91
4   Fas-Mediated Apoptosis 92
4.1   Apoptosis In Vitro 92
4.2   Apoptosis In Vivo 92
4.3   Activation of Fas to Induce Apoptotic Signal 94
4.4   Apoptotic Signal Mediated by Fas 94
5   Fas Ligand, a Death Factor 95
5.1   Identification and Purification of Fas Ligand 95
5.2   Molecular Properties of the Fas Ligand 96
5.3   Expression of the Fas Ligand 97
6   Physiological Roles of the Fas System 97
6.1   Involvement of the Fas System
        in Development of T-Cells...................... 97
6.2   Involvement of the Fas System
        in CTL-Mediated Cytotoxicity 98
6.3   Pathological Tissue Damage
        Caused by the Fas System 98
7   Perspectives 100
References 100

Myc-Mediated Apoptosis 104
Y. KUCHINO, A. ASAI, and C. KITANAKA

Abstract 104
1   Introduction 104
2   c-Myc-Mediated Apoptosis 105
3   s-Myc-Mediated Apoptosis 108
3.1   Structural Feature of the s-*myc* Gene 108
3.2   Structural Feature and Biological Functions
        of the *s-Myc* Protein 110
3.3   Expression of the *s-myc* Gene 114
3.4   *s-Myc*-Mediated Apoptosis 119
4   Escape from Myc-Induced Apoptotic Cell Death 123
5   Conclusions 124
References 126

Clusterin: A Role in Cell Survival in the Face of Apoptosis? ... 130
C. KOCH-BRANDT and C. MORGANS

Abstract 130
1   Introduction 130
2   Clusterin–a Widely Expressed Multifunctional Protein   132

3   Clusterin Gene Expression
in Apoptotic Epithelial Tissues   135
4   Clusterin Gene Expression in the Thymus   137
5   Clusterin Gene Expression
in Retinitis Pigmentosa (RP) Retina   138
6   Clusterin Expression During Degenerative Processes
in the Brain   139
7   Conclusions and Perspectives   143
References   144

List of Contributors

Addresses are given at the beginning of the respective contribution

Asai, A.   104
Cheney, C.M.   13
Dobmeyer, T.   58
Hartke, J.K.   13
Kabelitz, D.   58
Kelve, M.   44
Kitanaka, C.   104
Koch-Brandt, C.   130
Kuchino, Y.   104
Morgans, C.   130
Müller, W.E.G.   44
Nagata, S.   87
Neil, J.C.   13

Oberg, H-H.   58
Pechhold, K.   58
Pergande, G.   44
Perovic, S.   44
Phipps, A.J.   13
Pohl, T.   58
Rojko, J.L.   13
Rossol, R.   58
Schleger, C.   44
Shiokawa, D.   1
Tanuma, S.   1
Tsujimoto, Y.   72
Ushijima, H.   44

An Endonuclease Responsible for Apoptosis

S. Tanuma[1,2] and D. Shiokawa[1]

1 Introduction

Apoptosis is a form of cell death that plays important roles in physiological and pathological phenomena as diverse as embryogenesis, metamorphosis, hemopoiesis, and carcinogenesis (Kerr et al. 1972; Wyllie 1980; Berger 1985; Carson et al. 1986; Ucker 1987; Smith et al. 1989; Dive and Hickman 1991; Groux et al. 1992; Cohen et al. 1992; Tanuma et al. 1993). Although apoptosis is thought to be a gene-directed cell death, how this suicide is regulated is still unknown. Apoptosis is characterized morphologically by cell shrinkage, nuclear collapse, and cell fragmentation, known as apoptotic bodies, which is accompanied by internucleosomal DNA fragmentation (Kerr et al. 1972; Wyllie 1980). The signal transduction and determination processes in apoptosis initiated by apoptotic signals are complex and dependent on cell types and states. However, the DNA fragmentation, which is suggested to be catalyzed by a constitutive endonuclease, is a crucial process common in apoptosis, irrespective of the initial stimulus.

Activation of this endonuclease may be a central mechanism leading to cell death. The identification of this endonuclease is, therefore, an important step toward understanding the regulatory mechanism of apoptosis. Previous studies suggested that a sustained increase in intracellular Ca^{2+} during apaptosis may activate an endogenous endonuclease that mediates DNA fragmentation (Cohen and Duke 1984; Wyllie et al. 1984; Arends et al. 1990; McConkey et al. 1990). The putative endonuclease responsible for apoptosis seems to be a Ca^{2+}-dependent enzyme, since Ca^{2+} chelators and calmodulin inhibitors can prevent both DNA fragmentation and cell death, and Ca^{2+} ionophores induce apoptosis. Furthermore, apoptosis is known to be inhibited by Zn^{2+} (Cohen and Duke 1984; Wyllie et al. 1984; Arends et al. 1990; McConkey et al. 1990). Thus, a specific Ca^{2+}-dependent endonuclease, which is inhibited by Zn^{2+}, may have function in the DNA fragmentation during apoptosis. One of the promising approaches for the identification of such apoptotic endonuclease is to determine the nature of DNA fragmentation at the cellular level and compare it with the properties of endonucleases purified from apoptotic cell nuclei.

[1]Department of Biochemistry, Faculty of Pharmaceutical Science, Science University of Tokyo, Shinjuku-ku, Tokyo, Japan
[2]Research Institute for Bioscience, Science University of Tokyo, Noda, Chiba, Japan

Here, we first analyzed the nature of DNA fragmentation in rat thymocytes and human promyelocytic leukemia (HL-60) induced to undergo apoptosis by x-ray irradiation, glucocorticoids, or actinomycin D. Second, we purified and characterized nuclear endonucleases capable of internucleosomal DNA cleavage from rat thymocytes. We provide evidence that isolated nuclei contain at least three species of endonucleases tentatively named DNase α, β, and γ. We also present the first finding that only DNase γ produced 3'-hydroxyl (OH) and 5'-phosphoryl (P) ends of internucleosomal DNA fragments. The DNase γ but not DNase α and β absolutely required both Ca^{2+} and Mg^{2+} for full activity and were inhibited by Zn^{2+}. The striking correlation between the mode of hydrolysis and the cleavage ends produced in the apoptotic thymocytes suggests that the DNase γ is responsible for apoptotic internucleosomal DNA fragmentation in rat thymocytes. The relationship between these DNases from rat thymocytes and endonucleases previously purified from several mammalian cells will be discussed.

2   Results

We exposed rat thymocytes and human promyelocytic leukemia (HL-60) cells to three different conditions reported to cause apoptosis; (1) irradiation with 10 Gy x-rays; (2) incubation in the presence of 10^{-7} M dexamethasone; (3) incubation in the presence of 1 µg/ml actinomycin D. These treatments triggered extensive DNA fragmentation prior to cell death (Fig. 1). The DNA fragmentation occurred after 3~4 h incubation; it was easily detected by agarose gel electrophoresis of cellular DNA, and was prevented in the presence of 0.8 mM Zn^{2+}. The characteristic ladder pattern of DNA fragments contrasts with the smear observed with DNA degradation in necrotic cells. The cleavage ends were analyzed by endolabeling methods. If DNA fragments leave free 3'-OH and 5'-P ends, the resulting nucleosome ladders should be detected by 3' end-labeling of the extracted DNA by terminal deoxynucleotidyl transferase (TdT) and [α-^{32}P]dCTP, and by 5' end-labeling by polynucleotide kinase and [γ-^{32}P]ATP only after pretreatment of the DNA with alkaline phosphatase, respectively. The resultant autoradiograms (Fig. 1a, b, c) revealed that in both cases of apoptosis, 3'-OH and 5'-P ends were produced in the fragment ends. These results provide conclusive evidence that the apoptosis induced under these conditions is catalyzed by an endonuclease capable of generating 3'-OH/5'-P cleavage ends of DNA chains. This is important information to identify the endonuclease responsible for apoptosis.

We next attempted to purify the responsible endonuclease from the rat thymocytes. The strategy for detection of the endonuclease activity that cleaves internucleosomal regions used HeLa S3 cell nuclei as substrates, since they contain little endogenous endonuclease activity. An endonuclease activity specific for the linker regions of chromatin would produce nucleosomal ladders detected in agarose gel electrophoresis. Endonuclease activities present in isolated nuclei from rat thymocytes were solubilized. Essentially complete solubilization of nuclear endonuclease activities were obtained when the high ionic strength

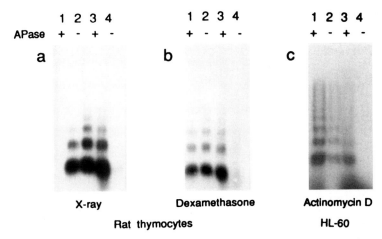

Fig. 1a–c. The modes of DNA fragmentations in apoptotic cells. The nature of DNA fragmentation was analyzed by end-labeling methods. DNA was extracted from apoptotic rat thymocytes treated with x-ray irradiation (**a**) or dexamethasone (**b**) and actinomycin D-treated HL-60 (**c**). The DNA was pretreated with (*lanes 1* and *3*) or without (*lanes 2* and *4*) alkaline phosphatase prior to 3'end- (*lanes 1* and *2*) or 5'end- (*lanes 3* and *4*) labeling. Aliquots of DNA were subjected to 2% agarose gel electrophoresis and autoradiography

sonication step was included. The soluble enzyme preparation was used in the subsequent chromatography. Three putative endonuclease species (DNase α, β, γ) were resolved in the third step of HPLC on CM5PW (Fig. 2).These DNase activities could catalyze the cleavage of linker DNA of chromatin in HeLa S3 cell nuclei and also cleave closed-circular plasmid DNA endonucleolytically. The induction of apoptosis by irradiation or dexamethasone resulted in a decrease in the activities of DNase α and β, whereas DNase γ activity was relatively constant. These observations were also seen in the apoptotic thymocytes induced by dexamethasone. Each activity was further purified by sequential HPLC steps on heparin5PW, G2000SW gel filtration and 2nd heparin5PW. We found no evidence for dissociable complexes of a single DNase or interconvertibility of the various forms in all chromatographies used. These enzyme preparations were used for studies of physiological and catalytic properties.

The molecular masses of these three DNases were determined by SDS-PAGE-renaturation system in gels containing double-stranded DNA. As shown in Fig. 3 upper panel, the localization of DNases within gels can be detected by disappearance of DNA flourescence as dark bands on an ethidium bromide-flourescent background. This activity gel method is based on the ability of DNase to renature after removal of SDS and to cleave DNA during incubation. Addition of Ca^{2+} and Mg^{2+} to the incubation medium with DNase γ, a non-flourescent band in the portion corresponding to a 33-kDa protein was seen. Both DNase α and β, after incubation under optimum conditions, exhibited nonflourescent bands corresponding to a protein of 32 kDa. These activities detected in the gel were correlated with those in DNA fragmentation assay (Fig. 3, lower panel). On

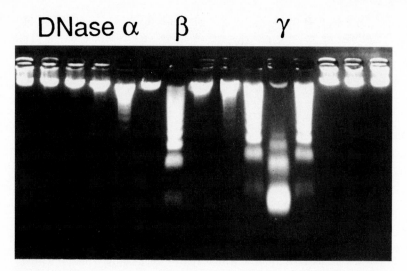

Fig. 2. CM5PW HPLC of nuclear DNases. The active fractions of DNases in DEAE5PW HPLC from apoptotic rat thymocytes were applied to CM5PW HPLC and eluted with a linear gradient of KCl (*left to right*). DNase activities of fractions were measured by HeLa S3 nuclear assay on agarose gel electrophoresis

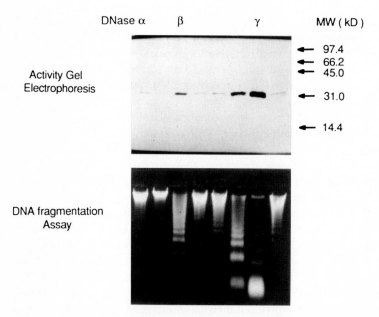

Fig. 3. Activity gel analysis of DNase α, β and γ. The molecular masses of DNase α, β, and γ were analyzed by activity gels on SDS-PAGE containing dsDNA (*upper panel*). The molecular mass protein markers were phosphorylase b (97 400), BSA (66 200), ovalbmin (45 000), carbonicanhydrase (31 000), soybean trypsin inhibitor (21 500), and lysoyme (14 400). DNase activities were also assayed by HeLa S3 nuclear assay (*lower panel*)

G2000SW gel filtration HPLC, DNase α, β, and γ appeared as a single peak of 28, 30, and 31 kDa, respectively, suggesting that these enzymes are monomeric polypeptides.

The mode of action of DNase γ is restricted to produce 3'-OH/5'-P ends. To determine this possibility, endolabeling methods were used (Fig. 4). The same labeling patterns as seen in Fig. 1a, b, c were observed, when the purified DNA from HeLa S3 cell nuclei digested with DNase γ were end-labeled at 3' ends with TdT and [α^{32}P] dCTP (Fig. 4c). The 5' ends of the fragments could not be labeled without alkaline phosphatase pretreatment. Thus, the DNase γ produced 3'-OH/5'-P ends of DNA chains. These labeling patterns were consistent with those seen in apoptotic rat thymocytes. In contrast, the DNA fragments formed by DNase α and β terminated with 3'-P/5'-OH ends, as evidenced by their ability to be labeled 3' ends by TdT only after alkaline phosphatase pretreatment, whereas 5' end fill-in reactions occurred without alkaline phosphatase pretreatment (Fig. 4a,b). Thus, the DNase γ present in apoptotic rat thymocyte nuclei is considered to be involved in DNA fragmentation during thymic apoptosis.

The optimum pH of DNase α, β, and γ in a HeLa S3 nuclear assay system was measured with various buffer systems. Optimum pH of both DNase α and β was around 5.6 in acetate-KOH or Mes-NaOH buffer. The DNase γ had a neutral pH optimum in the range of 6.8–7.8 with a maximum at pH 7.2 in Mops-NaOH buffer (Table 1). The DNase γ required both Ca^{2+} and Mg^{2+} for full activity. The presence of Ca^{2+} and Mg^{2+} together exerted a syneargistic effect on DNase γ activity (Table 1). The optimal concentrations for both were 1–3 mM. Among several other divalent cations tested, only Mn^{2+} was able to partially substitute for Ca^{2+}/Mg^{2+}. Interestingly, the DNase γ was sensitive to Zn^{2+}: a half-maximal

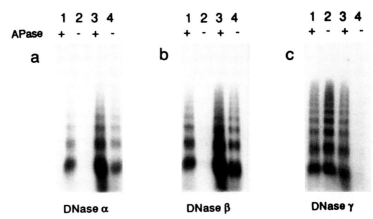

Fig. 4a–c. The modes of cleavage of DNase α, β, and γ. DNA was extracted from HeLa S3 cell nuclei digested by DNase α (**a**), β (**b**), and γ (**c**) and then subjected to 3' end- (*lanes 1* and *3*) or 5' end- (*lanes 3* and *4*) labeling after pretreatment with (*lanes 1* and *3*) or without (*lanes 2* and *4*) alkaline phosphatase. Aliquots of DNA were analyzed by 2% agarose gel electrophoresis and autoradiography

Table 1. Properties of DNase α, β, and γ in rat thymocytes

	Localization	Molecular mass	Optimal pH	Divalent cation requirement	Inhibition by Zn^{2+}	Mode of action
Dnase α	Nuclei	28 000[a] 32 000[b]	5.6	No	$IC_{50} > 1$ mM	endo- (3'-P, 5'-OH)
DNase β	Nuclei	30 000[a] 32 000[b]	5.6	No	$IC_{50} > 1$ mM	endo- (3'-P, 5'-OH)
DNase γ	Nuclei	31 000[a] 33 000[b]	7.2	Ca^{2+}/Mg^{2+}, Mn^{2+}	$IC_{50} = 40$ μM	endo- (3'-OH, 5'-P)

[a]TSKG 2000 SW gel filtration.
[b]SDS-PAGE.

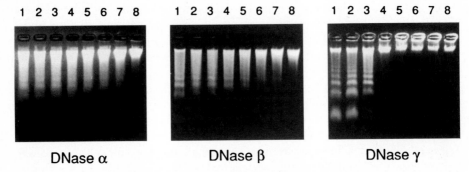

Fig. 5. Effect of Zn^{2+} ion on DNase α, β, and γ. The purified DNase α, β, and γ were assayed in the presence of 3 mM $MgCl_2$ and 3 mM $CaCl_2$ with varying concentrations of Zn^{2+}. Concentrations of Zn^{2+} were varied: 0, 0.01, 0.03, 0.1, 0.3, 1, 3, and 10 mM (*lanes 1–8*)

inhibition was achieved by Zn^{2+} concentrations as low as 40 μM (Fig. 5), which is about one order of magnitude lower than the extracellular concentration for inhibition of both DNA fragmentation and cell death. In contrast, the DNase α and β activities had essentially no effect on these divalent cations.

3 Discussion

A number of preparations of endonucleases have been purified from eukaryotic cells (Table 2). Judging from the physiological and catalytic properties of the purified DNase α, β, and γ, they are distinct from the well-known mammalian endonucleases, DNase I, II, V, and VI (Lascowski 1971; Dulaney and Touster 1972; Pedrini et al. 1976; Wang and Furth 1977; Anai et al. 1981; Lacks 1981; Kreuder et al. 1984; Liao 1985; Barry and Eastman 1993). These properties also distinguish them from previously purified mammalian endonucleases that require only one divalent cation, usually Mg^{2+} (Nakayama et al. 1981; Nagae et al. 1982; Hibino et al. 1988). Endonucleases that required both Ca^{2+} and Mg^{2+} have been purified from several mammalian sources (Table 2) (Hashida et al. 1982; Gaido

Table 2. Properties of Mammalian Endonucleases

DNase	Animal (tissue)	Localization	Molecular mass Native	Molecular mass SDS/PAGE	Optimal pH	Divalent cation requirement	Inhibitor	Reference
DNase I	Rat (parotid gland) bovine (pancreas)	Extracellular ER		32 000 31 000	$7.0{\sim}8.0^a$ 5.5^a	Ca^{2+}/Mg^{2+}, Mg^{2+}, Mn^{2+}	G-actin	Lascowski (1971) Lacks (1981) Anai et al. (1981) Kreuder et al. (1984)
DNase II	Rat (liver) Porcine (spleen)	Lysosomes	37 000 45 000 (heterodimer)	35 000 10 000	4.1	No	Iodoacetic acid, SO_4^{2-}	Dulaney and Touster (1972) Liao (1985)
DNase V	Calf (thymus)	n.d.	53 000 (tetramer)	13 000	6.6	Mg^{2+}, Mn^{2+}	n.d.	Wang and Furth (1977)
DNase VI	Calf (thymus)	n.d.	45 000	n.d.	9.5	Mg^{2+}, Mn^{2+}	n.d.	Pedrini et al. (1976)
Ca^{2+}/Mg^{2+}-endonuclease	Bull (seminal plasma)	Extracellular	28 000	36 000	$7.4{\sim}8.0$	Ca^{2+}/Mg^{2+}	Spermin N-ethylmaleimide Zn^{2+}	Hashida et al. (1982)
Nuc18	Rat (thymus)	Nuclei	n.d. Zn^{2+} n.d.	18 000	$7.0{\sim}8.5$	Ca^{2+}/Mg^{2+}	Zn^{2+} ATA^b	Gaido and Cidlowski (1991)
Ca^{2+}/Mg^{2+}-endonuclease	Human (spleen)	Nuclei	n.d.	27 000	8.0	Ca^{2+}/Mg^{2+}	ATA Spermin Zn^{2+}	Ribeiro and Carson (1993)
Ca^{2+}/Mn^{2+}-endonuclease	Rat (thymus)	Nuclei	22 000	22 000	$6.0{\sim}7.5$	Ca^{2+}/Mn^{2+}, Mn^{2+}	Iodoacetamide N-ethylmaleimide	Nikonova et al. (1993)
Mg^{2+} dependent endonuclease	Rat (liver)	Nuclei	36 000	36 500	n.d.	Mg^{2+}	Ca^{2+}	Hibino et al. (1988)

S. Tanuma and D. Shiokawa

(Table 2.) Contd.

DNase	Animal (tissue)	Localization	Molecular mass Native	Molecular mass SDS/PAGE	Optimal pH	Divalent cation requirement	Inhibitor	Reference
Neutral DNase	Rat (intestinal mucosa)	n.d.	36 000	n.d.	6.2^c / 7.7	Co^{2+}, Mn^{2+}	G-actin	Nagae et al. (1982)
Nuclear endonuclease	Bovine (small intestinal mucosa)	Nuclei	49 000 (heterodimer)	30 000 / 23 000	5.4	Mn^{2+}, Co^{2+}, Mg^{2+}	N-ethylmaleimide p-chloromercuri-benzoate	Nakayama et al. (1981)
CHO acidic endonuclease (DNaseII)	Chinese[d] hamster (ovary)	Nuclei	38 000	30 000 / 31 000 (doublet)	5.0~5.5	No	ATA Iodoacetic acid n-bromosuccinimide	Barry and Eastman (1993)

n.d., not determined.

[a]The upper and lower values were obtained in the presence of Ca^{2+}/Mg^{2+} and Mg^{2+} alone, respectively.

[b]Aurintricarboxylic acid.

[c]The upper and lower values were obtained in the presence of Co^{2+} and Mn^{2+}, respectively.

[d]The Chinese hamster ovary cells are a cultured cell line.

Table 3. Putative endonucleases for apoptosis in mammalian cells

DNase	Animal (tissue)	Localization	Moleculer mass	Optimal pH	Divalent cation requirement	Inhibition by			Mode of action	Reference
						Zn^{2+} (1mM)	G-actin (100 µg/ml)	ATA[d] (1 mM)		
Nucl8	Rat (thymus)	Nuclei	18 000	7.0~8.5	Ca^{2+}/Mg^{2+}	+[c]	n.d.	+	Endo	Gaido and Cidlowski (1991)
DNaseI	Bovine (pancreas)	Extracellular ER	31 000	7.0~8.0[b] 5.5	Ca^{2+}/Mg^{2+} Mg^{2+}, Mn^{2+}	−	+	−	Endo (3'-OH, 5'-P)	Peitsch et al. (1993)
CHO acidic endonuclease (DNaseII)	Chinese[a] hamster (ovary)	Nuclei	30 000 31 000 (doublet)	5.0~5.5	No	−	n.d.	+	Endo	Barry and Eastman (1993)
Ca^{2+}/Mg^{2+}-endonuclease	Human (spleen)	Nuclei	27 000	8.0	Ca^{2+}/Mg^{2+}	+ (IC_{50}=15 µM)	−	+	Endo	Ribeiro and Carson (1993)
Ca^{2+}/Mn^{2+}-endonuclease	Rat (thymus)	Nuclei	22 000	6.0~7.5	Ca^{2+}/Mn^{2+}, Mn^{2+}	n.d.	n.d.	n.d.	Endo	Nikonova et al. (1993)
DNase γ	Rat (thymus)	Nuclei	33 000	7.2	Ca^{2+}/Mg^{2+}, Mn^{2+}	+ (IC_{50}=40 µM)	−	+	Endo (3'-OH, 5'-P)	Shiokawa et al. (1994) Tanuma and Shiokawa (1994)

n.d., not determined.
[a] The Chinese hamster ovary cells are a cultured cell line.
[b] The upper and lower values were obtained in the presence of Ca^{2+}/Mg^{2+} and Mg^{2+} alone, respectively.
[c] 100% inhibition was attained at 2 mM $ZnCl_2$.
[d] Aurintricarboxylic acid.

and Cidlowski 1991; Ribeiro and Carson 1993; Nikonova et al. 1993). Their properties, however, differ between these studies and our DNase α, β, and γ (Table 1). Thus, we conclude that these DNase α, β, and γ are novel endonucleases that are present in nuclei and able to hydrolyze chromatin into nucleosomal oligomers. The presence of similar DNases in calf thymocytes (unpubl. data) suggests their general role in thymic nuclear function.

A common molecular event in thymic apoptosis is inter-nucleosomal cleavage of nuclear DNA, which is suggested to be induced by activation of a constitutive endogenous endonuclease (Cohen and Duke 1984; Wyllie et al. 1984; Arends et al. 1990; McConkey et al. 1990). The varying effects of Ca^{2+} and Zn^{2+} on apoptosis of rat thymocytes and the cleavage ends of DNA fragments can now be strikingly related to the selective action of DNase γ (Tanuma and Shiokawa 1994; Shiokawa et al. 1994). The properties distinguished it from purified Nuc18 (Gaido and Cidlowski 1991), Ca^{2+}/Mg^{2+}-(Reibeiro and Carson 1993), and Ca^{2+}/Mn^{2+}-(Nikonova et al. 1993) dependent endonucleases, which were recently reported to be putative apoptosis endonucleases (Table 3). These results suggest that DNase γ present in apoptotic thymocyte nuclei catalyzes DNA fragmentation during thymic apoptosis.

Although the physiological significance of the cognate form of DNase α and β is at present unknown, the specific activities of these DNases that are different in tissues (unpubl. data) provide some clues as to their functions *in vivo*. Definition of the essentiality of DNase γ in apoptosis *in vivo* and the distinct DNA fragmentation roles between these three DNases awaits further studies. Whether these DNases are products of separate genes or posttranslational modifications of a single gene product also remains to be determined. Further studies will be required to elucidate the activation mechanisms of these three nuclear DNases and their gene regulation. Such information may provide important clues for understanding the biological functions of nuclear DNases in cell death or survival.

Acknowledgments. This work was supported in part by a Grant-in-Aid for Scientific Research from the Ministry of Education, Science, and Culture of Japan, and in part through funds provided by the Nito Foundation.

References

Anai M, Sasaki M, Muta A, Miyagawa T (1981) Purification and properties of a neutral endo-deoxiribonuclease from guinea pig epidermis. Biochem Biophys Acta 656: 183–188
Arends MJ, Morris RG, Wyllie AH (1990) Apoptosis. The role of the endonuclease. Am J Pathol 136: 593–608
Barry MA, Eastman A (1993) Identification of deoxyribonuclease II as an endonuclease involved in apoptosis. Arch Biochem Biophys 300: 440–450
Carson DA, Seto S, Wasson DB, Carrera CJ (1986) Lymphocyte disfunction after DNA damage by toxic oxygen. J Exp Med 163: 746–751
Cohen JJ, Duke RC (1984) Glucocorticoid activation of a calcium-dependent endonuclease in thymocyte nuclei leads to cell death. J Immunol 132: 38–42

Cohen JJ, Duke RC, Fadok UA, Sellins KS (1992) Apoptosis and programmed cell death in immunity. Annu Res Immunol 10: 267–293

Dive C, Hickman (1991) Drug-target interactions: only the first step in the commitment to a programmed cell death? Br J Cancer 64: 192–196

Dulaney JT, Touster O (1972) Isolation of deoxyribonuclease II from rat liver lysosomes. J Biol Chem 247: 1424–1432

Gaido ML, Cidlowski JA (1991) Identification, purification, and characterization of a calcium-dependent endonuclease (NUC18) from apoptotic rat thymocytes. J Biol Chem 266: 18580–18585

Groux H, Torpier G, Monte D, Monton Y, Capron A, Ameisen JC (1992) Activation-induced death by apoptosis in CD4+ T cells from human immunodeficiency virus-infected asymptomatic Individuals. J Exp Med 175: 331–340

Hashida T, Tanaka Y, Matsunami N, Yoshihara K, Kamiya T, Tanigawa Y, Koide SS (1982) Purification and properties of bull seminal plasma Ca^{2+}, Mg^{2+}-dependent endonuclease. J Biol Chem 257: 13114–13119

Hibino Y, Yoneda T, Sugano N (1988) Purification and properties of a magnesium-dependent endodeoxyribonuclease endogenous to rat-liver nuclei. Biochem Biophys Acta 950: 313–320

Kerr JFR, Wyllie AH, Currie AR (1972) Apoptosis: a basic biological phenomenon with wide-ranging implications in tissue kinetics. Br J Cancer 26: 239–257

Kreuder V, Dieckhoff J, Sitting M, Mannherz HG (1984) Isolation, characterization and crystallization of deoxiribonuclease I from bovine and rat parotid gland and its interaction with rabbit skeletal muscle actin. Eur J Biochem 139: 389–400

Lacks SA (1981) Deoxiribonuclease I in mammalian tissues. Specificity of source. J Biol Chem 256: 2644–2648

Liao TH (1985) The subunit structure and active site sequence of porcine spleen deoxyribonuclease. J Biol Chem 260: 10708–10713

McConkey DJ, Hartzell P, Orreniuss (1990) Rapid turnover of endogenous endonuclease activity in thymocytes: effects of inhibitors of macromolecular synthesis. Arch Biochem Biophys 278: 284–287

Nagae S, Nakayama J, Nakano I, Anai M (1982) Purification and properties of a neutral endodeoxyribonuclease from rat small intestinal mucosa. Biochemistry 21: 1339–1344

Nakayama J, Fujiyoshi T, Nakamura M, Anai M (1981) Purification and properties of an endodeoxy-ribonuclease from nuclei of bovine small intestinal mucosa. J Biol Chem 256: 1636–1642

Nikonova LV, Beletsky IP, Umansky SR (1993) Properties of some nuclear nucleases of rat thymocytes and their changes in radiation-induced apoptosis. Eur J Biochem 215: 893–901

Pedrini AM, Ranzani G, Pedrali Noy GCF, Spadari S, Falaschi A (1976) A novel endonuclease of human cells specific for single-stranded DNA. Eur J Biochem 70: 275–283

Peitsch MC, Polzar B, Stepham H, Crompton T, MacDonald HR, Mannherz HG, Tshopp J (1993) Characterization of the endogenous deoxyribonuclease involved in nuclear DNA degradation during apoptosis (programmed cell death). EMBO J 12: 371–377

Ribeiro JM, Carson DA (1993) Ca^{2+}/Mg^{2+}-dependent endonuclease from human spleen: purification, properties, and role in apoptosis. Biochemistry 32: 9129–9136

Shiokawa D, Ohyama H, Yamada T, Takahashi K, Tanuma S (1994) Identification of an endonuclease responsible for apoptosis in rat thymocytes. Eur J Biochem 226: 23–30

Smith CA, Williams GT, Kingston R, Jenkinson EJ, Owen JJ (1989) Antibodies to CD3/T-cell receptor complex induce death by apoptosis in immature T cells in thymic cultures. Nature 337: 181–184

Tanuma S, Shiokawa D (1994) Multiple forms of nuclear deoxyribonuclease in rat thymocytes. Biochem Biophys Res Commun 203: 789–797

Tanuma S, Shiokawa D, Tanimoto Y, Ikekita M, Sakagami H, Takeda M, Fukuda S, Kochi M (1993) Benzylideneascorbate induces apoptosis in L929 tumor cells. Biochem Biophys Res Commun 194: 29–35

Ucker DS (1987) Cytotoxic T lymphocytes and glucocorticoid activate an endogenous suicide process in target cells. Nature 327: 62–64

Wang E, Furth JJ (1977) Mammalian endonuclease, DNase V. Purification and properties of enzyme of calf thymus. J Biol chem 252: 116–124

Wyllie AH (1980) Glucocorticoid-induced thymocyte apoptosis is associated with endogenous endonuclease activation. Nature 284: 555–556

Wyllie AH, Morris RG, Smith AL, Dunlop D (1984) Chromatin cleavage in apoptosis: association with condensed chromatin morphology and dependence on macromolecular synthesis. J Pathol 142: 67–77

Cytopathic Feline Leukemia Viruses Cause Apoptosis in Hemolymphatic Cells

J.L. Rojko[1,2,3], J.R. Hartke[1], C.M. Cheney[1], A.J. Phipps[1], and J.C. Neil[4]

Abstract

Certain isolates of the oncoretrovirus feline leukemia virus (FeLV) are strongly cytopathic for hemolymphatic cells. A major cytopathicity determinant is encoded by the SU envelope glycoprotein gp70. Isolates with subgroup C SU gp70 genes specifically induce apoptosis in hemolymphatic cells but not fibroblasts. In vitro exposure of feline T-cells to FeLV-C leads first to productive viral replication, next to virus-induced cell agglutination, and lastly to apogenesis. This in vitro phenomenon may explain the severe progressive thymic atrophy and erythroid aplasia which follow viremic FeLV-C infection in vivo. Inappropriate apoptosis induction has also been hypothesized to explain the severe thymicolymphoid atrophy and progressive immune deterioration associated with isolates of FeLV containing variant envelope genes. The influence of envelope hypervariability (variable regions 1 [Vr1] and 5 [Vr5] on virus tropism, viremia induction, neutralizing antibody development and cytopathicity is discussed. Certain potentially cytopathic elements in FeLV SU gp70 Vr5 may derive from replication-defective, poorly expressed, endogenous FeLVs. Other more highly conserved regions in FeLV TM envelope p15E may also influence apoptosis induction.

1 Introduction

Direct killing of hemolymphatic cells is central to the pathogenesis of certain immunosuppressive and myelosuppressive human and feline retrovirus infections. Basic questions about the relationship between retrovirus infection and hemolymphatic cell destruction are readily studied in in vitro assays, but not readily approached in retrovirus-infected human patients. Several laboratories, including ours, have been studying the cytopathic effects of the feline leukemia

[1] Department of Veterinary Pathobiology, The Ohio State University, Columbus, OH, 43210, USA
[2] Comprehensive Cancer Center, The Ohio State University, Columbus, OH 43210, USA
[3] Center for Retrovirus Research, The Ohio State University, Columbus, OH 43210, USA
[4] Department of Veterinary Pathology and Beatson Institute for Cancer Research, University of Glasgow, Bearsden, Glasgow G63 1QH, UK

oncoretroviruses (FeLVs) to determine their suitability as animal models for cytopathic human retrovirus infections. It is the intent of this article to discuss the evidence for cytopathicity in FeLV infections in vivo and in vitro, the apoptotic and nonapoptotic mechanisms used by FeLV to effect lymphoid and myeloid suppression, and the virogenes that may direct killing and suppression.

FeLV is a group of contagiously transmitted retroviruses which collectively account for more feline disease than any other single agent (reviewed in Rojko and Hardy 1993). One to two percent of the approximately 55 million American cats have chronic productive FeLV infections, with a circulating virus burden averaging 10^4 to 10^6 infectious virions/ml of plasma, and are termed viremic. Serosurveys indicate that between 20 and 40% of "sick" cats presented to veterinarians have chronic FeLV viremia. An additional 3 to 40% of American cats have been exposed to FeLV but have recovered from viremia; the incidence increases with urban environment and exposure to other cats in a multicat household. Some FeLV-exposed, recovered (regressor) cats remain latently infected and harbor a variably repressed or sequestered form of FeLV in their bone marrow, lymph nodes, spleen, and possibly other organs. Depending upon geographic region and other factors, between 10 and 40% of American cats are vaccinated against FeLV, using one of several commercial inactivated virion or subunit vaccines.

Eighty-three percent of chronically viremic pet cats succumb to FeLV-related diseases within 3.5 years of initial diagnosis of viremia (McClelland et al. 1980). About half of these cats die from T-cell immunosuppressive disorders and associated opportunistic infections, an additional fourth die of severe non-regenerative anemia (Francis et al. 1977, 1979; McClelland et al. 1980). Most of the rest die from other cytopathic diseases including neutropenias, enteritides with crypt cell destruction, wasting, and diminished growth. Only 5 to 10% of FeLV-viremic cats die of malignant neoplasms such as lymphomas, myelo-proliferative disorders, and fibrosarcomas. Most of these diseases have been reproduced by experimental inoculation of specific-pathogen-free cats by field isolates and biologic or molecular clones of FeLV (Jarrett et al. 1964; Hoover et al. 1972, 1973, 1974, 1987; Mackey et al. 1975; Jarrett and Russell 1978; Sarma et al. 1978; Onions et al. 1982; Riedel et al. 1986).

It is important to recognize that FeLV, like many oncoretroviruses, has a specific tropism for rapidly dividing cells like lymphoid, hematopoietic, and intestinal crypt cells and basilar epithelia in general (Rojko et al. 1978, 1979). The integration of the FeLV provirus into these mitotic cell populations provides not only for persistent or latent infection but also can serve as a insertional mutagen. As an integrated provirus, FeLV can deregulate or transduce oncogenes like *myc* which are critical to the regulation of cell proliferation (Levy et al. 1984; Mullins et al. 1984; Neil et al. 1984; Forrest et al. 1987). FeLV also has been shown to transduce the T-cell receptor *beta*-chain gene, a component of the cell surface complex vital to the control of immune cell proliferation (Fulton et al. 1987). Certain FeLV envelope surface unit SU glycoproteins and hydrophobic TM proteins directly induce severe cytostasis in hemolymphatic cells (Hartke 1992;

Rojko et al. 1992). Taken together, it is clear that FeLV/host cell interactions can lead to either neoplasia or senescence of mitotic cell populations critical for growth and maintenance of the organism.

The long prodromal period before disease development in nature results from the fact that only a minority of FeLVs are directly cytopathic (reviewed in Rojko and Hardy 1994). The FeLVs are readily subclassified by interference and antibody neutralization assays into three major subgroups: A, B, and C (Sarma and Log 1973). Several virus variants also exist which have been proposed to form either a variant A virus or a subgroup D virus (Kristal et al. 1993). These subclassifications are not moot. Studies conducted over the past 20 years indicate that subgroup specificity and biologic behavior including host cell and histogenetic tropisms, cytopathicity in vitro and pathogenicity in vivo are linked first and foremost to the viral surface-unit (external) envelope (*env*) glycoprotein gp70. FeLV-A is found in every infected cat in nature, is the only subgroup known to be transmitted contagiously, but by itself is virtually apathogenic (Sarma and Log 1973; Sarma et al. 1978; Jarrett and Russell 1978; Jarrett et al. 1973, 1978, 1984; Donahue et al. 1988). FeLV-B arises de novo in cats infected with FeLV-A and is the result of recombination of the integrated FeLV-A provirus with sequences highly related to FeLV endogenous to the cat genome, termed enFeLVs, which have been highly conserved during the last 1 to 10 million years of feline evolution (Benveniste et al. 1975; Elder and Mullins 1983; Soe et al. 1983, 1985; Nunberg et al. 1984; Stewart et al. 1986). FeLV-Bs replicate poorly in cats but are associated with the development of lymphomas and myeloproliferative diseases (Jarrett et al. 1978; Tzavaras et al. 1990; McDougall et al. 1994). FeLV-Cs and the variant FeLVs also emerge de novo from FeLV-A by mutation and in some cases, recombination (Rigby et al. 1992; Brojatsch et al. 1992). FeLV-C and the variant FeLVs are cytopathic; their appearance is generally the harbinger of acute fatal thymicolymphoid depletion (FeLV-C and variant FeLVs) and erythroid aplasia (FeLV-C) (Onions et al. 1982; Jarrett et al. 1984; Mullins et al. 1986, 1989, 1991; Hoover et al. 1987; Overbaugh et al. 1988, 1992; Dornsife et al. 1989a,b). FeLV-C and the variant FeLV also readily kill or induce cytostasis in feline T4-cells in vitro and in vivo (Donahue et al. 1991; Overbaugh et al. 1992; Rojko et al. 1992). Furthermore, cytopathic FeLVs display target cell speificity. When one examines the hemolymphatic tissues of chronically viremic pet cats for FeLV antigens indicative of productive infection, it appears that the majority of the bone marrow (essentially all cell types), the majority of the peripheral blood leukocytes and platelets, and the follicular and medullary areas of the lymph node are heavily infected (Hardy et al. 1973 a,b; Rojko et al. 1978, 1979, 1981; Dean et al. 1992) Also, certain FeLVs replicate readily in T-cells, B-cells, and monocytes in vitro (Hoover et al. 1980, 1981; Rojko et al. 1981; Grant et al. 1984). Yet, the targets for FeLV destruction appear to principally be T-cells and erythroid progenitors, although degeneration or destruction of intestinal crypt epithelia, neutrophils, and probably neural cells sometimes occurs. This discussion will focus primarily on the contributions of FeLV-induced apoptosis to the spectrum of spontaneous and experimental cytopathic disease in the cat and explore differences in virogenes of noncytopatic FeLV-A and cytopathic FeLV-C and variant subgroups most likely responsible for the differences in biologic behavior.

2   Cytopathic Infections with FeLV-C In Vivo and Direct Induction of Apoptosis in T4 Cells by FeLV-C

FeLV-C is rare in nature and is isolated from only 1% of viremic pet cats (Sarma and Log 1973; Jarrett and Russell 1978; Jarrett et al. 1978). However, FeLV-C infection is invariably associated with a rapid onset, fatal anemia, and often is associated with severe thymic atrophy and paracortical lymphoid depletion, circulating T and B lymphopenia, and other hemolymphatic deficits (Mackey et al. 1975; Cockerell et al. 1976; Onions et al. 1982; Riedel et al. 1986; Dornsife et al. 1989a). The pathogenesis of FeLV-induced thymic atrophy and nodal depletion is controversial. Some authorities find evidence for FeLV replication and lymphocyte ablation in early lesions (Anderson et al. 1971). Other investigators find minimal evidence for FeLV replication in the thymus or nodal paracortex and minimal evidence of thymic or paracortical lymphoid necrosis (Hoover et al. 1973; Rojko et al. 1979). T-cell depletion has been ascribed to impaired traffic of thymocyte precursors from marrow to thymus and impaired traffic of mature thymocytes to periarteriolar and perifollicular regions in the spleen and lymph nodes (Hoover et al. 1973; Rojko et al. 1979) or to FeLV-induced apoptosis (Hartke 1992; Rojko et al. 1992).

The prototype FeLV-C is FeLV-/Sarma, a virus originally biologically cloned from a field isolate, Kawakami-Theilen strain FeLV (FeLV-KT), which contains FeLVs of subgroups A, B, and C. Unlike most viruses, which kill cells by causing necrosis, FeLV-C/Sarma kills feline T-cells by inducing death by forcing physiologic maturation in the process variably called programmed cell death, terminal cell differentiation, or apoptosis (McCabe and Orrenius 1992; Rojko et al. 1992). Apoptosis is the process used by many cells which depend upon maturation for acquisition of function; such cells include lymphocytes, erythrocytes, enterocytes, and keratinocytes (reviewed in Duvall and Wyllie 1986). Hematopoietic growth factors expand the proliferating pool of progenitor cells in part by their repression of apoptosis (Williams et al. 1990). As detailed elsewhere in this book, apoptosis is recognized morphologically by the induction of nucleolar and nuclear condensation with eventual pycnosis and in the case of keratinocytes and erythrocytes, nuclear extrusion. A calcium-dependent endonuclease causes cleavage of the DNA fragments into internucleosomal fragments in multiples of approximately 200 base pairs. This results in a characteristic laddering pattern seen by analysis of fragment size on agarose gels (Duvall and Wyllie 1986; Wyllie 1987). Apoptosis in lymphocytes follows cytotoxic T-lymphocyte-target cell interaction, X-irradiation, chemotherapy, glucocorticoid treatment, and oxidative or thermal stress and perhaps clonal deletion and induction of tolerance, but its induction by viruses is rare but may occur following HIV infection (Sellins and Cohen 1987; Ucker 1987; Kisielow et al. 1988; Tomei et al. 1988; Yamada and Ohyama 1988; Smith et al.

1989; Liu and Janeway 1990; Laurent-Crawford et al. 1991; Meyaard et al. 1992).

Specifically, in vitro inoculation of feline CD4-positive, CD8-positive lymphoma 3201 cells with biologic or molecular clones of FeLV-C/Sarma or its parental FelV-KT leads to a acute onset, highly productive infection, down-regulation of surface CD4 with persistence of surface CD8, and rosette-forming capability indicative of maturation and death of the cells within 7 to 12 days (Hartke 1992; Rigby et al. 1992; Rojko et al. 1992; Rojko JL and Cheney CM, unpubl.). Inoculation of 3201 cells with FeLV-A leads to a gradual onset, moderately productive, infection without noticeable death. While both FeLV-A and FeLV-C are replicated rapidly and to high titer by feline fibroblastoid FEA cells, neither FeLV-A nor FeLV-C kill FEA cells. FeLV-C killing of feline T4-cells is via induction of apoptosis. Hours to days prior to death, FeLV-C-infected cells cluster together tightly and demonstrate nuclear and nucleolar condensation, surface blebbing, and fragmentation (Figs. 1, 2). DNA fragmentation and laddering occur 1 to 2 days before massive cell death (Fig. 3). Other antemortem changes include a shift from phospholipid to neutral lipid incorporation of [^{14}C]-oleic acid, increases in palmitic acid proportions and decreases in linoleic and proportions (Rojko et al. 1992). Increased palmitate can directly kill cells and potentiate tumor necrosis factor and tumor necrosis factor can cause apoptosis or necrosis in its target cells (Bjerve et al. 1987). It will be useful to determine if T4-cells from FeLV-viremic cats shows similar fatty acid shifts. Plasma from viremic cats show decreases in arachidonic acid and increases in palmitoleic acid proportions (Williams et al. 1993). Others have shown that FeLV-C-infected fibroblastoid cells also can initiate the release of tumor necrosis factor from uninfected feline bone marrow cells, feline peritoneal macrophages, and human histiocytic lymphoma U937 cells (Khan et al. 1992 a,b,c, 1993).

3 Killing of Lymphocytes by FeLV-FAIDS/p61C and Associated Variants

A highly pathogenic variant of FeLV designated FeLV-p61C or FeLV-FAIDS (for feline acquired immunodeficiency syndrome) was isolated originally from the intestine of a viremic pet cat with thymic lymphoma and severe immunodeficiency (Mullins et al. 1986; Hoover et al. 1987). Inoculation of cats with FeLV-FAIDS field isolate led to rapid onset fatal immunodeficiency and enteritis. Histologic lesions in necropsied cats included severe widespread thymic atrophy and paracortical lymphoid depletion and destruction of intestinal crypt epithelia, lesions very reminiscent of acute radiation-induced damage which may have an apoptotic mechanism (Mullins et al. 1986; Hoover et al. 1987).

Cloning and sequence analysis of the FeLV-FAIDS/p61C virus complex have revealed a contagiously transmitted, replication-competent, noncytopathic FeLV-A designated FeLV-A/p61E. It has also revealed nine replication-defective variants, six of which could be rescued as transmissible virus, following cotransfection of feline cells with FeLV-A/p61E (Overbaugh et al. 1988, 1992). Three of these were originally cloned from small intestine DNA and have been designated

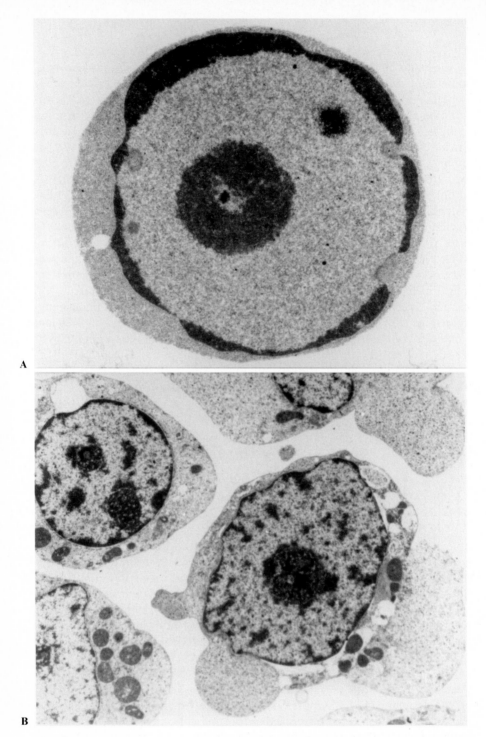

Fig. 1 A, B. Transmission electron micrographs of FeLV-C/Sarma-infected 3201 cells, 8 days after inoculation. There are increasing nuclear and nucleolar condensation (**A,B**), blebbing of the cytoplasmic membrane (**A**), and morphologically normal mitochondria (**A**). (Photograph courtesy Rojko et al. (1992)

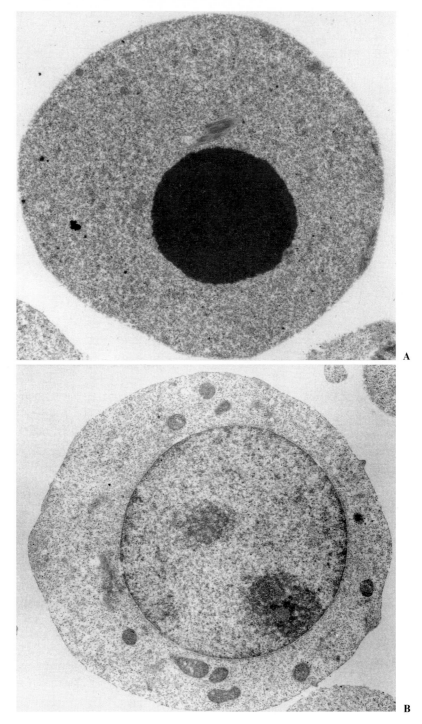

A

B

Fig. 2A, B. Transmission electron micrographs of FeLV-C/Sarma-infected 3201 cell (**A**) and sham-inoculated 3201 cell (**B**), 8 days after inoculation. Overt pycnosis is evident in FeLV-C/Sarma-infected cell (**A**) while the sham-inoculated cells contains finely dispersed chromatin, two prominent nucleoli, scant cytoplasmic organelles, morphologically normal mitochondria, and an intact, smoothly rounded cytoplasmic membrane (**B**). (Photograph courtesy Rojko et al. 1992)

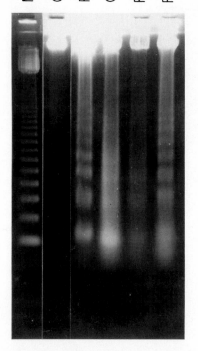

Fig. 3. Ethidium bromide-strained agarose gel demonstrating fragmentation of 10 µg DNA from 3201 cells from FeLV-C/Sarma-inoculated cultures 4 days postinoculation. Minimal laddering of DNA from sham-inoculated cultures at 1, 4, and 5 days postinoculation and in FeLV-C/Sarma-infected cultures at 1 and 5 days postinoculation. (Photograph courtesy Rojko et al. 1992)

p61B, p61C, and p61D. Another three were cloned from bone marrow DNA and designated 61-1, 82K, and 45H. When inoculated in conjunction with replication-competent FeLV-A/61E and tested for the ability to kill feline T4-lymphoma 3201 cells, variants 61C, 61-1, 82K, 61B, and 45H were cytopathic; however, variant 61B consistently took 5 to 7 days longer to kill (Donahue et al. 1991; Overbaugh et al. 1992). When inoculated by itself, FeLV-A/61E did not replicate well in 3201 cells and was not cytopathic (Donahue et al. 1991; Overbaugh et al. 1992). Cytologic and molecular changes seen in FeLV-FAIDS/p61C-inoculated feline T-lymphoma 3201 cells were compatible with death via apoptosis induction (Donahue et al. 1991; Mullins JI, pers. comm. 1991).

In serial infections with the FeLV-FAIDS/p61C complex (e.g., cat passage), it is clear that the common form (e.g., FeLV-A/61E) replicates faster than the variant form(s) but that disease occurs only after the variant form(s) appear in the bone marrow (Mullins et al. 1986, 1989, 1991; Hoover et al. 1987; Overbaugh et al. 1988, 1992). In the acute form of FAIDS (mean survival time of 3 months), there is a short prodromal period of 2 to 10 weeks in which only the common FeLV-A form is replicated. The onset of immunodeficiency follows the emergence of tissue-specific variant p61C genome replication, accumulation of integrated and unintegrated variant viruses in high copy number, and cell death in bone marrow, intestine and lymph node. In chronic FAIDS (mean survival time greater than 1 year), the prodrome lasts 6 months to 1 year and precedes the appearance of other variant genomes with substantial deletions in *env*, *pol*, and elsewhere, and the development of severe immunodeficiency. An early lymphoid hyperplasia associated with the replication of FeLV in lymphoid follicles precedes eventual T cell ablation in blood, thymus, and nodes, loss of lymphocyte blastogenic responses and ability to reject cutaneous allografts, hypogamma-globulinemia, intractable diarrhea resulting from intestinal crypt cell death with or without secondary opportunists, and sometimes bacterial respiratory disease, and nonregenerative anemia. Surviving cats often develop lymphoma (Hoover et al. 1987; Mullins et al. 1986, 1989, 1991; Overbaugh et al. 1988, 1992; Quackenbush et al. 1989, 1990). It has been shown that all the variant FeLV genomes found in chronic FAIDS existed in the initial inoculum and that differential replication during serial cat passage reflects the selection of preexist-ing mutants and not rapid *de novo* evolution of new variants (Mullins et al. 1986, 1989, 1991; Hoover et al. 1987; Overbaugh et al. 1988, 1992). Again, though, the onset of clinical disease usually follows the emergence of the variant virus genomes and the accumulation of unintegrated viral DNA in bone marrow, lymph node, and intestine.

p61C is replication-defective because it has deletions in its structural group-associated antigen (*gag*) and polymerase (reverse transcriptase, *pol*) and enve-lope (*env*) genes (Overbaugh et al. 1988). When inoculated into cats in conjunction with replication-competent FeLV-p61E, FeLV-p61C directly de-stroys the lymph node and bone marrow cells in which it is replicated. Specifically, FeLV-p61C proviral DNA is produced to excess and these excess FeLV proviruses cannot be integrated into cat bone marrow or lymph node DNA. Accumulation of excess, unintegrated FeLV DNA in the cytoplasm of cells appears to be very toxic. Nodal and marrow T lymphocytes are eliminated and profound immunosuppression ensues. Unintegrated retroviral DNA has also been implicated as toxic in the killing of cells by avian retroviruses and mammalian lentiviruses (including HIV) in vivo and in vitro (Weller et al. 1980; Shaw et al. 1984; Fisher et al. 1986; Haase 1986) but not in cytopathic infections with FeLV-C/Sarma (Mullins et al. 1986).

The cytopathic effect of FeLV-FAIDS/p61C also is linked to an incomplete, delayed, posttranslational processing of glucose residues in the viral SU *env* glycoprotein. These aberrantly processed *env* SU glycoprotein molecules fail to migrate to the plasma membrane to become part of the budding virus and rather accumulate in the cytoplasm where they may disrupt normal functions or other-wise kill the cell (Poss et al. 1989, 1990).

4   FeLV-C-Related Nonregenerative Anemia: Potential Role for Apoptosis

Nearly 70% of all nonregenerative anemias in cats in nature are the result of
FeLV infection (Hardy et al. 1973 a,b; Cotter et al. 1975; Essex et al. 1975;
Mackey et al. 1975; Hardy 1981; Kociba 1986; Rojko and Hardy 1994). The
packed cell volume (PCV, a measure of the circulating red cell mass) may fall to
4 to 14% as compared with normal values for age-matched cohorts of 25 to 35%.
Hemoglobin levels may be as low as 1 to 4 g/dl. Erythrocyte numbers may fall
to 1.5 to 3×10^6/ml. The bone marrow is often hypocellular due to the disappearance of erythroid progenitors and viremic anemic pet cats have relative
decreases in metarubricytes and rubricytes and may have relative increases in
prorubricytes and rubriblasts. The residual prorubricytes and rubriblasts show
cytoplasmic vacuolation and mild megaloblastic changes. Persistence of
rubriblasts and prorubricytes has been interpreted to indicate maturation
arrest. This is not likely as even earlier erythroid progenitors called burst-
forming units-erythroid (BFU-E) and colony-forming units-erythroid
(CFU-E) are expunged from the marrow of FeLV-positive, preanemic, and
anemic cats (see below). Serum erythropoietin levels are increased 23 times
relative to normal cats (321 versus 14 mU/ml by radioimmunoassay) and 6.4
times relative to phlebotomized cats (321 versus 49 mU/ml) (Kociba 1986;
Kociba 1986; Wardrop et al. 1986).

   The pathogenesis of erythroid aplasia probably is dictated by the two FeLV
env proteins, SU gp70 and TM p 15E. As discussed above, *env* gp70s are responsible for subgroup specificity (e.g., A versus B versus C). Subgroup C FeLV is
routinely isolated from viremic pet cats with erythroid aplasia (Jarrett et al. 1978;
Onions et al. 1982; Brojatsch et al. 1992; Rigby et al. 1992). Also, experimental
introduction of FeLV-C, particularly FeLV-C/Sarma, isolates into specific-
pathogen-free cats causes an acute onset, rapidly progressive, fatal, erythroid
aplasia (Hoover et al. 1974; Boyce et al. 1981; Kociba et al. 1982; Onions et al.
1982; Testa et al. 1983; Abkowitz et al. 1985, 1987a; Riedel et al. 1986, 1988;
Rigby et al. 1992) which coincides exactly with the appearance of FeLV-C in the
plasma (Jarrett et al. 1984) and with the expression of retroviral proteins by
erythroid progenitors (Abkowitz et al. 1987b) and accessory cells like macrophages (Kociba and Halper 1987).

   The crucial factors are FeLV-C infection of accessory macrophages and other
stromal cells and FeLV-C infection and destruction of two classes of erythroid
progenitor cells, BFU-E and CFU-E. BFU-E and CFU-E are identified by
in vitro assays for their growth and colony formation in semisolid medium.
BFU-E are less mature than CFU-E, both BFU-E and CFU-E are less mature
than rubriblasts. The earliest studies showed that BFU-E are infected first and
begin to disappear beginning about 10 days after exposure to FeLV-C infection
while CFU-E inhibition occurs 10 to 30 days later and marks the beginning of
fatal erythroid aplasia (Testa et al. 1983; Abkowitz et al. 1987 a, b). Other
investigators have demonstrated loss of CFU-E prior to loss of BFU-E but
concomitant with a loss of the ability of BFU-E to respond to growth factors

(Abkowitz 1991). Myelomonocytic progenitor cells, designated colony-forming units granulocyte/macrophage, usually are not altered by FeLV-C infection (Riedel et al. 1986, 1988; Abkowitz et al. 1987a,b; Rigby et al. 1992).

When Geoffroy X domestic cat F_1 hybrid cats are exposed to FeLV-C, maturation from BFU-E to CFU-E is arrested (Abkowitz et al. 1985; Abkowitz et al. 1987 a,b). What actually happens is this. First, FeLV-C infection increases the fraction of BFU-E in DNA synthesis causing an early increased cycling of BFU-E. Next, the BFU-E can no longer respond to hematopoietic growth factors, and the CFU-E disappear, the packed cell volume falls rapidly, and the cat develops a fatal nonregenerative anemia. Whether similar events occur in viremic pet cats has not been determined. In experimentally infected cats, however, this failure to respond to hematopoietic growth factors may implicate an apoptotic mechanism when one considers that hematopoietic growth factors, particular erythropoietin, keep cells cycling in part by their inhibition of the apoptotic program (Williams et al. 1990).

The bone marrow microenvironment also is pivotal in the development of idiopathic aplastic anemias in man, in retroviral marrow suppression in primates, and in FeLV-induced erythroid aplasias in cats (Watanabe et al. 1990; Steinberg et al. 1991). The marrow microenvironment is composed of stromal cells like fibroblasts, macrophages, adventitial reticular cells, endothelial cells, and adipocytes (Liesveld et al. 1989). The marrow microenvironment also contains the extracellular matrix and its attachment protein fibronectin which provides sites for hematopoietic cell anchoring. Marrow accessory cells and stromal fibroblasts and abnormal in both human aplastic anemias (Tsai et al. 1987) and in FeLV-C-induced erythroid aplasias in cats (Wellman and Kociba 1988; Wellman et al. 1988; Linenberger and Abkowitz 1992 a,b). Cats viremic with FeLV-KT (subgroups ABC) have decreased numbers of marrow stromal cell progenitors called colony-forming units-fibroblast (CFU-F) and FeLV-KT and FeLV-C both replicate in marrow fibroblasts and marrow accessory macrophages (Wellman et al. 1988; Dornsife et al. 1989b; Linenberger and Abkowitz 1992 a,b; Khan et al. 1992a,b,c, 1993). Replication of FeLV in bone marrow accessory cells can trigger the release of tumor necrosis factor-*alpha* which can depress CFU-F, BFU-E, and CFU-F (Khan et al. 1993), particularly when interferon-*gamma* also is present (Khan et al. 1992b,c). In vitro, FeLV infection of feline fibroblast FEA cells increases the production of multilineage, hematopoietic growth factors normally produced constitutively by FEA cells. These growth factors act directly on marrow progenitors to increase output of erythroid, myelomonocytic, and magakaryocytic series cells (Abkowitz et al. 1986; Linenberger and Abkowitz 1992b). FeLV infection of marrow stromal cells may also provide a reservoir of infected cells, thereby increasing the likelihood of cell-associated transfer of virus from infected to unifected cells which appears to be important in the development of FeLV-C infections in vitro and probably in vivo (Dornsife et al. 1989a,b).

5   FeLV Genes Important in Cytopathicity/Apoptosis Induction

5.1   General Remarks

Despite the wide differences between FeLV-A and FeLV-C in terms of biologic behavior in vivo and in vitro, comparison of the sequences of the molecular clones of FeLV-A and FeLV-C currently available shows that FeLV-A and FeLV-C differ very little (being about 98% homologous) but that the differences are clustered in five variable regions in the structural *env* SU gp70 (Vr1 through Vr5, see Fig. 4) and to a lesser extent in the long terminal repeat (LTR) virogenes that control transcription and thus influence replication.

5.2   T-Cell Killing

The subgroup-specific T-cell killing by FeLV probably does not result solely from differences in LTR-controlled replicative ability in critical target cells. It is clear that avirulent FeLV-A (Glasgow 1 to 61E) isolates replicate poorly in T-cells in vitro compared to cytopathic FeLV-C/Sarma or FeLV-FAIDS/p61C variant isolates (Table 1) and that this replicative efficiency may reflect the transcriptional inefficiency of the FeLV-A LTR in T-cells compared to that of the FeLV-C or FeLV-FAIDS/p61C LTR in T-cells (Donahue et al. 1991; Overbaugh et al. 1992; Rojko et al. 1992). These differences in LTR transcriptional activity and productivity of retroviral infection are not seen in fibroblasts which resist FeLV-C and FeLV-FAIDS/p61C cytopathicity. The LTRS of cytopathic and noncytopathic FeLVs differ by very little (less than 2 to 5%) (Riedel et al. 1986; Stewart et al. 1986; Overbaugh et al. 1988, 1992; Fulton et al. 1991). Notably, however, the regulatory U3 region of the FeLV-C LTR contains the majority of the 21 point mutations and several of these point mutations are in the enhancer region, with 2 in the putative glucocorticoid-response element, 2 in the Nuclear Factor 1 binding site, and one at the site of binding of an as yet-unidentified nuclear factor *flvi-1* implicated in splenic nonT-, nonB- lymphomagenesis (Riedel et al. 1986; Stewart et al. 1986; Levesque et al. 1990; Fulton et al. 1990). Relative to the FeLV-A LTR, the p61C LTR shows even less change, as only two point mutations in the U3 enhancer regions and one point mutation at the U3/R junction have been demonstrated (Overbaugh et al. 1988; Fulton et al. 1990). Despite these differences, avirulent FeLV-A isolates replicate as well as or better than the cytopathic FeLV-C isolates in cat lymphocytes in vivo (Hamilton 1984; Dean et al. 1992). Rather, it is the current consensus that the dominant virogenes responsible for pathogenicity in vivo and T-cell killing in vitro reside in the variable regions of the *env* SU gp70.

The region of greatest difference between FeLV-C and FeLV-A is Vr1 in the N-terminus of gp70 (Fig. 4). Modeling of the *env* SU glycoprotein genes of a variety of mammalian and avian retroviruses predicts that Vr1 should encode hydrophilic virion surface-associated determinants that specify receptor binding

Table 1. Envelope GP70 sequences of FeLV isolates

FeLV	Sequence at Vr1	Sequence at Vr5	Replication in 3201 cells	Cytopathicity in 3201 cells
	PIVLNPTNVKHGARYSSSKYG	LCNKTQQGHTGAH YLAAPN		
A/Glasgow-1	PIVLNPTNVKHGARYSSSKYG	LCNKTQQGHTGAH YLAAPN	Slow	No
A/61E	. . . . . S	. . . .	Slow	No
C/Sarma	. . APD . . . . RSW . . . . P .	. K . . K . . K . T .	Fast	Apoptosis
AC/Sarma	. . APD . . . . RSW . . . . TH .		Slow	No
var p61C	. M . . S . . . . G . PP	. . . . . . H . . . . DYLTAPR	Fast[a]	Yes
var p61B	. M . . S . . . G . PP	. . . . . . H . . . . DYLAAPR	Fast[a]	Yes
var 82K	. M . . S . . . G . PP	. . . . . . H . . . . DYRAAPR	Fast[a]	Yes

[a]Variants p61C p61B and 82K are defective for replication and require coinfection with helper replication-competent FeLV-A/61E.

<START

```
              MESPTHPKPS KOKTLSWNLA FLVGILFTID IGMANPSPHQ IYWVTWITN VQTNTQANAT SMLGTLTDAY PTLHVDLCDL VGDTWEPIVL NPTNVKHGAR YSSS  KYGCKT  TDRKKQQQTY PFVCPGHAP
A-Glasgow     .......... .......... .......... .......... ......... .......... .......... .......... .......... .......... ....  ......  .......... .........
A-1161E       .......... .......... ....V..... .......... ......... .......... .......... .......... .......... .......... ....  ......  .......... .........N
p61C          .......... .......... ....V..... .......P..M .......... .......... .......... .......... ..M..S.... ........G. .PP.  ......  .......... .N
C-SARMA       .......... .......... .FP....... ......Q..M. ........V. .......SR. .......... .......... ....AP..D. ..RSW..... .TH.  ......  .......... .N
C-Z215        .......... .......... .......... .......V... ........V. .......SR. .......... .......... ....MAP.D. ..RSW..... .TH.  ......  .......... .N
C-FA27        Not done
C-FS246       Not done
B-RICKARD     Not done   .......... .......... .........T .......V.. .......... ......F..Y ....MYF... ..I...N..P .DL.WGW..S ..S.  .....DQ ..PM..RW.. .RNT.....A
B-GA          .......... ......V... .V.LRL.... .........T .......V.. .........Y ......F... ....MYF... ..I...N... ...S.RP..S ...SP .....DQ ..PM..RW.. .RNT.....N
B-ST          .......G.. .F.D.M..I. .V.LRL.A.. .........I .......TF. ........V. ......TF.. ....IYF... ..I...N... .........S ..SDQ EPFPG ..PM..RW.. .RNT.....N
             .......G.. .F.D.I..I. .V.LRL.V.. .........I .......V.I .......T.. ......F... ....YF.... ..I...N... .........S ..SDQ EPF.PG ..PM..RW.. .RNT.....N
EndogCFE6                                                                                                                       ..SDQ EPF.PG ..PM..RW.. .RNT.....A
EndogCFE16                                                                                                                      ..SDQ EPF.PG ..PM..RW.. .RNT.....N
```

Vr1

```
              SSQDNS CE GKCNPLVL QFTQKGRQAS WDGPKMWGLR LYRTGYDPIA LFTVSRQVST
A-Glasgow     ...... .. ........ .......... .......... .......... ..........
A-1161E       ...N.. .. .I...... .......K.. .......... .......... ..........
1161C         ...N.. .. .I...... .......K.. .......R.. ........S. .......M..
C-SARMA       ...N.. .N .K...... .......... .......R.. ........S. .......M..
C-Z215        ...N.. .N .K...... .......... .......... .......... ..........
C-FA27        ...... .P ........ .......... .......... .......... ..........
C-FS246       ...... .. ........ .......... .......... .......... ..........
B-RICKARD     ...... .. ..TTGA.G.G.R. .......I.. .......S.. ........S. .......M..
B-GA          ...... .. ..ITGA.S.G.R. .......I.. .......S.. ........S. .......M..
B-ST          ...... .. ..ITGA.S.G.R. .......I.. .......S.. ........S. .......M..
EndogCFE6     ...... .. ..ITGA.S.G.R. .......I.. .......S.. ........S. .......M..
EndogCFE16    ...... .. ..ITGA.S.G.R. .......I.. .......S.. ........S. .......M..
```

Vr3

```
              SLGPKGTHCG GAQDGFCAAW GCETTGETWW KPTSSWDYIT VKRG
A-Glasgow     .......... .......... .......... .......... ....
A-1161E       .......... .......... ....A..S.. .......... ....
1161C         .......... .......... ....A..S.. .......... ....
C-SARMA       .M....Y... .......... ....A...T. .......... ....
C-Z215        .M....Y... .......... .......T.. .......... ....
C-FA27        .......... .......... .......... .......... ....
C-FS246       .......... .......... .......... .......... ....
B-RICKARD     RKQ....... P......... ......Y..R .......... ....
B-GA          RKQ....... .P........ ......Y..R .......... ....
B-ST          RKQ....... .P........ ......Y... .......... ....
EndogCFE6     RKQ....... .P........ ......Y... .......... ....
EndogCFE16    RKQ....... .P........ .......... .......... ....
```

Vr2

```
              ........ K.VTQGIY QCSGGGWCGP CYDKAVH.
A-Glasgow     ........ .K.VTQGIY QCSGGGWCGP CYDKAVH.
B-RICKARD     ........ .K.VTQGIY QCSGGGWCGP CYDKAVH.
B-GA          ........ .K.VTQGIY QCSGGGWCGP CYDKAVH.
B-ST          ........ .K.VTQGIY QCNGGGWCGP CYDKAVH.
EndogCFE6     ........ .K.VTQGIY QCNGGGWCGP CYDKAVH.
EndogCFE16    ........ .K.VTQGIY QCSGGGWCGP CYDKAVH.
```

```
              A PRSVAPTTMG PKRIGTGDRL INLVQGTYLA LNATDPNKTK DCWLCLVSRP PYYEGIAILG NYSNQTNPPP
A-Glasgow     . .......... ....V..... .......... .......... .......... .......... ..........
1161C         . .......... .......... .......... .......... .......... .......... ..........
C-SARMA       T .......SA. .......... .......... .......... .......... .......VS. ..........
C-Z215        T .......SA. .......... .......... .......... .......... .......VS. ..........
C-FA27        . .......I.. .......... .......... .......... .......... .......... ..........
C-FS246       . .......G.. .......... .......... .......... .......... .......... ..........
B-RICKARD     .IE.R.T.PHHS.G.GGT.PGITLVNASI. .L.TPV.PAS ........N. .......V.. .......... ..........
B-GA          .IE.R.T.PHHS.G.GGT.PGITLVNASI. .L.TPV.PAS ........... .V.N...... .......... ..........
B-ST          .IE.R.T.PHHS.G.GGT.PGITLVNASI. .L.TPV.PAS ........N. .V.N....R. .......... ..........
             .P .IE.R.I.PHPP.G.GGT.PGITLVNASI. .L.TPV.PAS ........N. .V.N...... ....T..... ..........
EndogCFE6     .D.P .IE.R.R.PHPS# (term.)
EndogCFE16
```

Vr4

```
                                                                         Vr5
          SCLSIPQHKL TISEVSGQGL CIGTVPKTHQ ALCNKTQQGH TGAH         YLAAPN GTYWACNTGL TPCISMAVLN WTSDFCVLIE LWPRVTYHQP EYVVTHFAKA VRFRR
A-Glasgow  .......... .......... .......... .......... ....                .......... .......... .......... .......... .......... .....
A-1161-E   ..I..P.... .......... ....H..... .......... DYLTAPR             .......... .......... .......... .......... .........G ....
1161C      ...T...... .......... .......... ..K..K..K.T ....               .......... .......... .......... ....I....D ......G...
C-SARMA    ...T...... .......... .......... ..K..K..K.T ....               .......... .......... .......... ....I....D Not done
C-Z215     .......... .......... .......... ..E....... ....                .......... .......... .......... .......... Not done
C-FA27     .......... .......... ....M..... .......... ....                .......... .......... .......... .......... Not done
C-FS246    ...D...... .......... .......... ..K..K..K.T ....         .V.S   .......... .......... .......... .......... .........
B-RICKARD  .......... .......... .......... ..K..K.T... ....                .......... .......... .......... .......... ....A.....
B-GA       ...D...... .......... .......... ..E....... ....         ..S    .......... I......... .......... .......... .........
B-ST       ...D...... .......... .......... ..K..K.T... ....                .......... I......... ....T..... ...D.T V.I .A........
EndogCFE6  ...V...... .......... ..A....... ..K..R.T... ....         .V.    .......... .......... ....T..... ..I.E..... I.S.ENK P.K
```

Fig. 4. Feline leukemia virus envelope sequences[a]

[a]Derived amino acid sequences and gp70 subregion identification compiled from Elder and Mullins (1983); Nunberg et al. (1984); Riedel et al. (1986); Stewart et al. (1986); Donahue et al. (1988); Overbaugh et al. (1988); Brojatsch et al. (1988); Rigby et al. (1992). Regions identified in order from 5' to 3' (e.g., left to right and top to bottom) include: Vr1 (major determinant of subgroup identity and host range, contributes to pathogenicity): Vr2 (no identified properties); Vr3 (no identified properties); Vr4 (no identified properties); Vr5 (determinant of T-cell killing).

[b]Regions identified by double lines above sequence indicate variable regions designated Vr1–Vr5 by Riedel et al. (1986, 1988).

and host range (Battini et al. 1992). Since FeLV subgroup specificity is determined *a priori* by interference assay which depends on blockade of homologous but not heterologous receptors on the cell surface (Sarma and Log 1973), by inference, Vr1 most likely should confer subgroup identity. Recent experiments by our laboratories and those of others have shown that this is indeed the case. Vr1 is the major and probably the sole determinant of subgroup A/C identity and host range and contributes to pathogenicity in vivo (Brojatsch et al. 1992; Rigby et al. 1992).

As discussed above, a replication-defective FeLV, FeLV-FAIDS/p61C, also kills T-cells when coinfected with helper FeLV-A/61E. As FeLV-FAIDS/p61C and FeLV-C have overlapping 6 and 3 codon deletions in the Vr1 region of gp70 relative to FeLV-A (Mullins et al. 1986; Riedel et al. 1986, 1988; Stewart et al. 1986; Donahue et al. 1988; Overbaugh et al. 1992), it has been suggested that the Vr1 region might be responsible for cytopathicity. However, experiments from our laboratories have clearly shown that recombinant FeLVs which contain a Vr1 region from cytopathic FeLV-C on an apathogenic FeLV-A background do not induce apoptosis or any form of cytopathicity in feline T-cells (Rigby et al. 1992). Thus, the Vr1 of FeLV-C/Sarma or FeLV-C/FZ215 is insufficient to confer T-cell cytopathicity (Table 1, Fig. 4; Rigby et al. 1992).

In addition, others have shown that the destructive capability of FeLV-FAIDS variant viruses resides in a C-terminal 34 amino acid segment of gp70 located at Vr5 where the p61C variant and other related cytopathic FAIDS variant viruses contain a short stretch of 6 inserted and substituted amino acids relative to FeLV-A (see Fig. 4, Table 1) (Poss et al. 1989, 1990; Donahue et al. 1991; Overbaugh et al. 1992). Although the specific viral determinants of apoptosis induction by FeLV-C/Sarma have not yet been identified, it is plausible that these determinants also reside in Vr5. Interestingly, not all FeLV-Cs are identical at Vr5 (Fig. 4). Two T-cell cytopathic isolates, FeLV-C/Sarma and FeLV-C/FZ215, are identical at Vr5, while two other isolates of undetermined cytopathicity, FeLV-C/FA27 and FeLV-V/FS246, resemble the horizontally transmitted, exogenous FeLV-As. It may also be worth noting that two FeLV-Bs, Rickard and Snyder-Theilen, contain Vr5 regions homologous to FeLV-C/Sarma and FeLV-C/FZ215 (Fig. 4). Rickard FeLV (FeLV-R) is associated with a severe preneoplastic thymic atrophy in vivo and Snyder-Theilen FeLV (FeLV-ST) is associated with fatal erythroid hypoplasia and thymic atrophy in vivo (Hoover et al. 1974; Hoover EA and Rojko JL, unpubl.). It should be very instructive to determine the in vitro T-cell cytopathic potential of these and other FeLVs which are homologous to, or different from, FeLV-C/Sarma and FeLV-C/FZ215 at Vr5.

The Vr5 region is of particular interest as a determinant of cytopathicity because of several considerations. First, molecular modeling predictions suggest that it resides on the surface of the *env* SU gp70 molecule and that the Vr5 region of FeLV-C/Sarma, FeLV-C/FZ215, FeLV-B/Rickard, and FeLV-B/Snyder-Theilen can potentially form a basic amphipathic *alpha*-helix with protruding lysine residues to engage the target cell surface (Hartke 1992). Short peptides with similar basic amphipathic *alpha*-helical potential have recently become recognized as important effectors of protein interactions such as calmodulin binding and as cytolysins and leucine zippers (Landshulz et al. 1988; Miller and Montelaro 1992; Miller et al. 1991, 1993a,b). Notably, synthetic peptide monomers homologous to codons 828–855 of the carboxy terminus of the *env* transmembrane (TM) gp41 of HIV can bind calmodulin and can directly lyse both bacterial and mammalian cells (Miller et al. 1993a,b). Although these peptides and the parental HIV-1 virus clearly induce the rapid, sustained influx of extracellular calcium and other small molecules into cells (Cloyd et al. 1989; Miller et al. 1993a,b), the HIV TM peptides do not appear to trigger apoptosis but rather broadly lyse a wide variety of cell types. A recent report also indicates that recombinants of FeLV-C with additional inserted enFeLV sequences cause aggregation of, and cytostasis in feline T-lymphoma 3201 cells; however, the sequence at Vr5 was not presented for these novel FeLV-C recombinants (Pandey et al. 1991).

A second reason for examining the role of Vr5 in the induction of apoptosis is that the Vr5 region most likely is expressed in vivo as cats that regress acute FeLV infection and become immune or latently infected develop antibodies to Vr5 demonstrable by enzyme-linked immunosorbent assay (Hartke 1992).

5.3 Erythroid Aplasias

The FeLV *env* TM protein p 15E adversely affects many sites in the cat's immune system; it also damages BFU-E and CFU-E. When uninfected cat bone marrow cells are exposed to infectious FeLV, inactivated FeLV, or purified p15E in vitro, BFU-E, CFU-E, and CFU-F are all suppressed; CFU-GM are usually not affected (Rojko et al. 1986; Wellman et al. 1984, 1988, 1991).

One of the difficulties with ascribing all the marrow depression associated with FeLV to p15E is that the p15E of FeLV-A is not very different from the p15E of FeLV-C. While some studies have indicated FeLV-C replicates much better than FeLV-A in bone marrow and peritoneal macrophages (Khan et al. 1993), others show equal replication of FeLV-A and FeLV-C in hematopoietic cells in vivo (Quackenbush et al. 1990; Dean et al. 1992). It seems more likely that the ability of FeLV-C to ablate marrow cells is (at least in part) determined by *env* SU gp70 sequences and it has been hypothesized that FeLV-C gp70 impairs BFU-E or CFU-E differentiation by interference with ligand/receptor interactions or signal transduction pathways unique to erythroid cells (Zack and Kociba 1988; Abkowitz 1991). However, FeLV-C and not FeLV-A also can induce apoptosis in feline thymocytes (Rojko et al. 1992). Although both FeLV-A and FeLV-C are able to induce production of multilineage hematopoietic growth factors from infected feline fibroblasts (Abkowitz et al. 1986; see above), only FeLV-C and not FeLV-A is able to induce tumor necrosis factor release from marrow accessory cells and tumor necrosis factor will initiate the apoptosis program (Bjerve et al. 1987; Khan et al. 1992a). Like thymocytes, which undergo apoptosis upon maturation, erythroid progenitors undergo apoptosis/programmed cell death when they differentiate into erythrocytes capable of transporting oxygen. Further investigations into a shared mechanism for apoptosis induction in both erythroid and lymphoid progenitors by FeLV-C gp70 should shed light on these phenomena (Hartke 1992). The homologous external *env* glycoprotein gp120 of HIV will directly induce tumor necrosis factor and inhibit human hematopoietic progenitors (Clouse et al. 1991). In other cases, infection with HIV may initiate the apoptotic program by an as yet undetermined mechanism (Laurent-Crawford et al. 1991; Groux et al. 1992; Meyaard et al. 1992). Also, certain modulators of cyclic 3', 5'-adenosine monophosphate (cAMP) and calcium can influence cellular apoptotic and differentiation pathways. It is interesting to note that bone marrow cells from FeLV-viremic and unifected cats react differently to treatment with pharmacologic agents, leading to enhancement of erythroid progenitors from normal cats but further inhibition of erythroid progenitors from viremic cats (Zack and Kociba 1988).

There may be more than one subregion of gp70 which determines the severity of anemia development. Early studies have linked the development of aplastic anemia to the N-terminal region of FeLV-C gp70 (Riedel et al. 1988). Furthermore, the Vr1 region which resides in this N-terminal region of gp70 is known to confer subgroup identity and host range in tissue culture assays (Rigby et al. 1992; Brojatsch et al. 1992). However, chimeric viruses which contain the Vr1 of FeLV-C inserted into an FeLV-A background and act as subgroup C viruses in vitro cause a Coombs'-negative, macrocytic anemia with partial regeneration, retention of CFU-E, and BFU-E, and intermittent reticulocytosis in vivo (Rigby et al. 1992). This type of anemia is somewhat reminiscent of the Coombs'-positive, macrocytic, partially regenerative (sometimes termed hemolytic) anemia seen in some viremic pet cats (Cotter et al. 1975; Weiser and Kociba 1983; Kociba 1986; see above) but does not resemble FeLV-C-induced erythroid aplasia. It seems reasonable that both Vr1 and another as yet undetermined region of FeLV-C gp70 (these authors would consider Vr5 a prime candidate) contribute to anemia development (see also Hartke 1992).

6   Speculative Relationship to Endogenous Virus Recombinants and Importance Across Evolution

Also of marked interest is the homology between the Vr5 region of the cytopathic FeLV-Cs with the endogenous FeLV (enFeLV) proviruses intrinsic to the cat genome. Recent studies by our laboratories and others have shown that enFeLv proviruses are widely expressed by feline hemolymphatic tissues in vivo and by feline lymphoid cells in vitro (McDougall et al. 1994). However, the transcribed sequences encode only the N-terminal half of the enFeLV proviral *env* gp 70 which is expressed and secreted from cells. One could speculate that expression of the full length enFeLV proviral gp70, particularly Vr5-related information, in hematopoietic or lymphoid tissues might lead to the demise of cells via premature differentiation/apoptosis.

7   Regions of FeLV p15E Important in Immunosuppression

Although the transmembrane *env* protein p15E is highly conserved between FeLV-A and FeLV-C, it is still plausible that p15E is involved in nonsubgroup-specific events that unbalance signal transduction and facilitate apoptosis induction in lymphoid or erythroid cells. Specifically, exposure of lymphocytes from cats, people, or mice with p15E or with ultraviolet-inactivated FeLV impairs T-cell responses and erythropoiesis in vitro (Mathes et al. 1978; Copelan et al. 1983; Cianciolo et al. 1985; Lewis et al. 1985; Orosz et al. 1985; Dezzutti et al. 1989). Various mechanisms have been documented including p15E-induced dysregulation of cytokines, protein kinase C, adenyl cyclase, and E-series prostaglandins (PGEs). Unbalanced signaling could lead to clonal deletion (via apoptosis) of retrovirus-specific T-cells (see below). The antiproliferative activity attributed to FeLV p15E appears to reside in a short peptide midway through the N-terminus of p15E which has been designated CKS-17 and has the following sequence: LQNRRGLDLLFLKEGGL (Cianciolo et al. 1985). The most important question about the antiproliferative activity of p15E remains to be answered: does p15E simply keep the cell inactive in a premitotic/relatively dormant phase of the cell cycle or does it force terminal cell differentiation/death via the apoptotic pathway?

8   Other Types of Cytopathic Disease Seen in Viremic Cats Which Could Have an Apoptotic Basis

8.1   Enteritis

A specific syndrome of FeLV enteritis is diagnosed on the basis of clinical, histopathologic, and virologic criteria (Reinacher 1987). In a recent study, 157 of 218 pet cats with chronic enteritis (72%) were viremic with FeLV (Reinacher 1987). FeLV has specific tropism for intestinal crypt cells and certain strains of FeLV seemingly are directly cytopathic for intestinal cells. Lesions at necropsy range from minimal catarrhal enteritis to extensive hemorrhagic or fibrinonecrotizing enteritis. Histologically, there are degeneration, necrosis, and dilatation of epithelial crypts. The remaining viable crypt epithelia is heavily infected with FeLV and viral destruction of crypt cells leads to cessation of viral replication. Some pet cats with severe acute cytopathic enteric disease also have evidence of myeloid and lymphoid destruction. This constellation of lesions also is seen in specific-pathogen-free cats inoculated as neonates or weanlings with the cytopathic FeLV-FAIDS/61C and sometimes occurs in specific-pathogen-free cats inoculated with the cytopathic FeLV-KT (subgroups A, B, C). It is interesting that the FeLV-FAIDS variant A/p61C virus actually was cloned originally from an intestinal lesion (Mullins et al. 1986; Hoover et al. 1987; Overbaugh et al. 1988, 1992). It seems reasonable to assume that the mitotic cell tropism of FeLV leads to the potential for mitotic cell destruction/senescence when the infecting virus is cytopathic; whether the pathogenesis involves apoptosis or necrosis has not been established. It is notable, however, that similar lesions are seen in radiation-induced enteritides and lymphopenias which are thought to have an apoptotic pathogenesis (Wyllie 1987; Tomei et al. 1988; Yamada and Ohyama 1988).

8.2 Infertility and Abortion

Historically, about two thirds of infertile or aborting queens (female cats) have been FeLV-viremic (Cotter et al. 1975; Hardy 1981). The syndrome of infertility can be reproduced by experimental transplacental FeLV-Rickard infection and is caused by fetal death, resorption, and placental involution in the middle trimester (Hoover et al. 1983). The precise pathogenesis of the reproductive failure is unknown but surely involves infection and destruction of fetal and placental cells by FeLV (Hardy 1981; Hoover et al. 1983). Infection of pregnant mice with certain murine leukemia viruses leads to fetal death and resorption. FeLV also will infect and reduce the viability of hamster fetuses, as shown in an embryo transfer study (Chapman et al. 1974).

8.3 Neurologic Syndrome

Both oncoretroviruses and lentiviruses have been associated with the development of neurologic disease in ungulates, mice, humans, and cats (Hardy 1981; Haase 1986; Gardner et al. 1973; Gardner 1985; Haffer et al. 1987; Bhavagati et al. 1988; Price et al. 1988; Ringler et al. 1988; Wheeler et al. 1990; Dow and Hoover 1992). While lentiviral infections principally appear to damage brain, spinal cord, and peripheral nerves, oncoretroviral infections are more likely to impair the spinal cord and peripheral nerves and spare the brain (Dow and Hoover 1992).

Lower motor neuron paralysis, locomotory and behavioral abnormalities, and sensory and motor polyneuropathy have all been described in FeLV-viremic cats (Haffer et al. 1987; Wheeler et al. 1990; Dow and Hoover 1992). Cats with experimental FeLV infections have abnormal spinal cord and peripheral nerve conduction velocities (Wheeler et al. 1990). FeLV also has been isolated from the cerebrospinal fluid of three cats with neurologic signs, but FeLV-specific antibodies was not detected in the cerebrospinal fluid (Dow and Hoover 1992). Hind limb paresis in Lake Casitas wild mice has been etiologically linked to the Cas-Br strain of murine leukemia virus (Gardner et al. 1973; Gardner 1985). The disease occurs in 2% of Lake Casitas wild mice older than 1 year of age. Degenerative lesions and found in the anterior lateral horns of the lumbosacral cord; they are characterized by spongiosis, gliosis, and neuronal vacuolation and resemble those found in amyotrophic lateral sclerosis in humans. MuLV virions are seen in the extracellular spaces of the neuropil and budding from neurons, endothelial cells, and glial cells (Gardner et al. 1973; Gardner 1985).

9 So How Does FeLV Cause Apoptosis?

It seems reasonable to suggest that FeLV p15E or an FeLV-C-specific viral protein in Vr5 or specific regions in the transmembrane *env* gp41 of HIV could impair signal transduction and lead to clonal deletion (via apoptosis) of retro-virus-specific T-cells. Ultraviolet-inactivated FeLV-KT (subgroups A, B and C) and the HIV transmembrance gp41 have been shown to impair protein kinase C activation and stimulate Ca^{2+} mobilization in T-cells, thereby unbalancing signal transduction as has been theorized for positive (activation) and negative (apoptosis) selection of thymic T-cell precursors (Duvall and Wyllie 1986; Dezzutti et al. 1989; McConkey et al. 1990; Ruegg and Strand 1990). It will be useful to determine which subregions of FeLV *env* SU gp70 or TM p15E are responsible for this unbalanced signaling (Hartke 1992). Inappropriate induction of a form of programmed cell death in T4-cells may occur in HIV infections (Ameisen and Capron 1991; Laurent-Crawford et al. 1991; Terai et al. 1991; Groux et al. 1992). Thus, FeLV immunosuppression could be mediated through activation of endonucleases in thymocytes unprotected by activated protein kinase C (Dezzutti et al. 1989; Rojko et al. 1992).

As apoptosis is death by forced differentiation, mature (e.g., already fully differentiated) cells like resting lymphocytes are somewhat protected from apoptosis. However, cortical thymocytes and antigen-stimulated lymphocytes proliferate and differentiate to acquire function. Cells undergoing activation or mitosis may be especially susceptible to inappropriately timed apoptosis induction by FeLV; certainly T-cell killing by HIV and SIV is facilitated by prior activation (Ameisen and Capron 1991; Laurent-Crawford et al. 1991; Rosenberg et al. 1991; Groux et al. 1992; Meyaard et al. 1992).

10 Other Retroviruses and Animal Viruses as Potential Causes of Apoptosis

10.1 Retroviruses that Behave as Superantigens

Superantigens are molecules that bind to certain T-cells via the variable region of the T-cell receptor-*beta* chain, thereby bypassing the requirement for T-cells to recognize T-dependent antigens in the context of the major histocompatibility complex (Janeway 1990). Both bacterial toxins (especially staphylococcal toxins) and (in mice) endogenous minor lymphocyte antigens (Mls) can act as super-antigens and activate or ablate portions of the T-cell repertoire. Remarkably, these minor lymphocyte antigens actually are encoded by endogenous murine mammary tumor retroviruses which have integrated into the germline as stable, heritable, DNA proviruses (Choi et al. 1991). This type of retroviral immune modulation may well be but the tip of the iceberg, as some evidence for specific clonal deletion as been presented for both MuLV and HIV (Cheung et al. 1991; Hugin et al. 1991; Imberti et al. 1991). That cytopathic FeLV isolates can behave as superantigens is plausible but has not yet been documented. It is particularly plausible given the clinical picture of selective but not generalized immunosuppression in FeLV infection (reviewed in Rojko and Hardy 1994) and the ability of cytopathic FeLV isolates to induce programmed cell death/terminal differentiation/apoptosis in certain T-cell subsets in vitro and thymic atrophy in vivo. Induction of apoptosis in certain thymic precursor cells could result in the absence of some but not all T-cells and potentially create holes in the T-cell repertoire and the observed selective acquired immunodeficiency in FeLV-viremic cats.

10.2 HIV and Other Lentiviruses

The induction of apoptosis in activated versus resting human hemolymphatic cells by HIV is discussed in detail elsewhere in this Volume. It is certainly reasonable to assume that hemolymphatic, neural, and other cell death in HIV infection may result from either apoptotic (Ameisen and Capron 1991; Laurent-Crawford et al. 1991; Terai et al. 1991; Meyaard et al. 1992), syncytium-inducing (Hildreth and Orentas 1989), necrotic, or directly lytic (Lynn et al. 1988; Cloyd et al. 1989; Miller and Montelaro 1992; Miller et al. 1991, 1993a,b) mechanisms either singly or in combination, depending on the virus strain, progression of the infection, status of the immune system, activation status or stage of the cell cycle, and a multiplicity of other factors.

It is increasingly probable that HIV-associated apogenesis contributes to altered thymic development in children exposed to HIV in utero (Papiernik et al. 1992; Bonyhadi et al. 1993). The relative contributions of widespread HIV replication in intrathymic T-cell precursors (Schnittman et al. 1990; Valentin et al. 1994) and altered microenvironmental structure and cytokine composition (Schnittman et al. 1991) warrant further study.

The immunosuppressive feline lentivirus, feline immunodeficiency virus (FIV) also appears to trigger apoptosis in vitro. Simultaneous exposure to FIV and tumor necrosis factor leads to killing of feline fibroblasts (Ohno et al. 1993), and FIV itself may kill feline T-cells in vitro by apoptosis induction (Tochikura et al. 1990; Bishop et al. 1993; Ohno et al. 1994) and promulgate lymphoid destruction in vivo. Whether this is the direct result of virally encoded products or secondary to FIV-induced cytokine dysfunction (Ohashi et al. 1992) remains to be determined.

The evidence for apoptosis induction in simian immunodeficiency virus (SIV) infections is controversial. Certain strains like the acutely pathogenic SIV-PBj may accelerate lymphoid destruction as well as proliferation in pigtail macaques (Fultz and Zack 1994). Apoptosis is increased in CD8-depleted peripheral blood lymphocytes of rhesus monkeys exposed to pathogenic SIV isolates (Gougeon et al. 1993; Estaquier et al. 1994). In other instances, SIV markedly alters thymic infrastructure but does not hasten thymocyte death (Muller et al. 1993). In a recent study, apoptotic lymphocytes were detected in the peripheral blood of monkeys with reactivated HIV-2 infections following SIVmac251/32H superinfection (Petry et al. 1995).

10.3 Chicken Anemia Virus

Chicken anemia virus, a small single-stranded DNA virus, causes anemia and thymocyte loss within 2 weeks after infection and is responsible for considerable economic losses in the poultry industry (Goryo et al. 1989). Recent studies indicate that chicken anemia virus induces DNA fragmentation and morphologic changes compatible with apoptosis in chicken thymocytes following in vivo and in vitro exposure. Whether apoptosis is an obligatory event in the viral life cycle (perhaps facilitating spread of virus as epithelial cells appear to phagocytose apoptotic thymocytes containing virus-like particles) or whether it is merely circumstantial warrants further study (Jeurissen et al. 1992). It is notable that chicken anemia virus and FeLV apparently share the same primary targets for cytopathicity: lymphoid and erythroid cells. Whether common mechanisms are employed by the two viruses has not been investigated.

Acknowledgments. This work was supported by grants AI 25722 (J.L.R.) and DK 41066 (J.L.R.), by NRSA 1 F32 CA60251-01 (J.R.H.) from the National Institutes of Health, and by the Cancer Research Campaign of the United Kingdom (J.C.N).

References

Abkowitz JL (1991) Retrovirus-induced feline pure red blood cell aplasia: Pathogenesis and response to suramin. Blood 77: 1442–1451

Abkowitz JL, Ott RL, Nakamura JM, Steinmann L, Fialkow PJ, Adamson JW (1985) Feline glucose-6-phosphate dehydrogenase cellular mosaicism. Application to the study of retrovirus-induced pure red cell aplasia. J Clin Invest 75: 133–140

Abkowitz JL, Holly RD, Segal GM, Adamson JW (1986) Multilineage, non-species specific hematopoietic growth factor(s) elaborated by a feline fibroblast cell line: enhancement by virus infection. J Cell Physiol 127: 189–196

Abkowitz JL, Holly RD, Adamson JW (1987a) Retrovirus-induced feline pure red cell aplasia: the kinetics of marrow failure. J Cell Physiol 132: 571–577

Abkowitz JL, Holly RD, Grant CK (1987b) Retrovirus-induced feline pure red cell aplasia: hematopoietic progenitors are infected with feline leukemia virus and erythroid burst forming cells are uniquely sensitive to heterologous complement. J Clin Inves 80: 1056–1063

Ameisen JC, Capron A (1991) Cell dysfunction and depletion in AIDS: the programmed cell death hypothesis. Immunol Today 12: 102–105

Anderson LJ, Jarrett WFH, Jarrett O, Laird HM (1971) Feline leukemia virus infection of kittens: mortality associated with atrophy of the thymus and lymphoid depletion. J Natl Cancer Inst 47: 807–817

Battini J-L, Heard JM, Danos O (1992) Receptor choice determinants in the envelope glyocproteins of amphotropic, xenotropic, and polytropic murine leukemia viruses. J Virol 66: 1468–1475

Benveniste RE, Sherr CJ, Todaro GJ (1975) Evolution of type C viral genes: origin of feline leukemia virus. Science 190: 886–888

Bhavagati S, Ehrlich G, Kula RW et al. (1988) Detection of human T-cell lymphoma/leukemia virus type 1 DNA and antigen in spinal fluid and blood is patients with chronic progressive myelopathy. N Engl J Med 318: 1141–1147

Bishop SA, Gruffydd-Jones TJ, Hardbour DA, Stokes CR (1993) Programmed cell death (apoptosis) as a mechanism of cell death in peripheral blood mononuclear cells from cats infected with feline immunodeficiency virus (FIV) Clin Exp Immunol 93(1): 65–71

Bjerve KS, Espevik T, Kildahl-Amdersen O, Nissen-Meyer J (1987) Effect of free fatty acids on the cytolytic activity of tumor necrosis factor/monocyte-derived cytotoxic factors. Acta Pathol Microbiol Immunol 15: 21–26

Bonyhadi M, Rabin L, Salimi S, Brown DA, Kosek J, McCune JM, Kaneshima H (1993) HIV induces thymic depletion in vivo. Nature 363: 728–732

Boyce JT, Hoover EA, Kociba GJ, Olsen RG (1981) Feline leukemia virus-induced erythroid aplasia: in vitro hemopoietic culture studies. Exp Haematol 9: 990–1001

Brojatsch J, Kristal BS, Viglianti GA, Khiroya R, Hoover EA, Mullins JI (1992) Feline leukemia virus subgroup C phenotype evolves through distinct alterations near the N-terminus of the envelope surface glycoprotein. Proc Natl Acad Sci USA 89: 8457–8461

Chapman AL, Weitlauf HM, Bopp W (1974) Effect of feline leukemia virus on transferred hamster fetuses. J Natl Cancer Inst 52: 583–586

Cheung SC, Chattopadhyay SK, Hartley JW, Morse HC III, Pitha PM (1991) Aberrant expression of cytokine genes in peritoneal macrophages from mice infected with LP-BM5-MuLV, a murine model of AIDS. J Immunol 146: 121–127

Choi Y, Kappler JW, Marrack P (1991) A superantigen encoded in the open reading frame of the 3' long terminal repeat of mouse mammary tumor virus. Nature 350: 203–207

Cianciolo GJ, Copeland TD, Oroszlan S, Snyderman R (1985) Inhibition of lymphocyte proliferation by a synthetic peptide homologous to retroviral envelope proteins. Science 230: 453–455

Clouse KA, Cosentino LM, Weih KA, Pyle SW, Robbins PB, Hochstein HC, Natarajan V, Farrar WL (1991) The HIV gp120 envelope protein has the intrinsic capacity to stimulate monokine secretion. J Immunol 147: 2892–2901

Cloyd MW, Lynn WS (1991) Perturbation of host cell membrane is a primary mechanism of HIV cytopathology. Virology 181: 500–511

Cloyd MW, Lynn WS, Ramsey K, Baron S (1989) Inhibition of human immunodeficiency virus (HIV-1) infection by diphenylhydentoin (dilantin) implicates role of cellular calcium in virus life cycle. Virology 173: 581–590

Cockerell GL, Krakowka S, Hoover EA (1976) Characterization of feline T- and B-lymphocytes and identification of an experimentally induced T-cell neoplasm in the cat. J Natl Cancer Inst 57: 1095–1102

Copelan E, Rinehart J, Lewis M, Mathes LE, Olsen RG, Sagone A (1983) The mechanisms of retrovirus suppression of human T-cell proliferation in vitro. J Immunol 131: 2017–2024

Cotter SM, Hardy WD Jr, Essex ME (1975) Association of feline leukemia virus with lympho-sarcoma and other disorders in the cat. J Am Vet Med Assoc 166: 449–454

Dean GA, Groshek PM, Mullins JI, Hoover EA (1992) Hematopoietic target cells of anemogenic subgroups C versus nonanemogenic subgroup A feline leukemia virus. J Virol 66: 5561–5568

Dezzutti CS, Wright KA, Lewis MG, Lafrado LF, Olsen RG (1989) FeLV-induced immuno-suppression through alterations is signal transduction: down regulation of protein kinase C. Vet Immunol Immunopathol 21: 55–67

Donahue PR, Hoover EA, Beltz GA, Riedel N, Hirsch VM, Overbaugh J, Mullins JI (1988) Strong sequence conservation among horizontally transmissible, minimally pathogenic, feline leukemia viruses. J Virol 62: 722–731

Donahue PR, Quackenbush SL, Gallo MV, deNoronha C, Overbaugh J, Hoover EA, Mullins JI (1991) Viral genetic determinants to T-cell killing and immunodeficiency disease induction by FeLV-FAIDS. J Virol 65: 4461–4469

Dornsife RE, Gasper PW, Mullins Jl, Hoover EA (1989a) Induction of aplastic anemia by intra-bone marrow inoculation of a molecularly cloned feline retrovirus. Leuk Res 13(9): 745–755

Dornsife RE, Gasper PW, Mullins Jl, Hoover EA (1989b) In vitro erythrocytopathic activity of an aplastic anemia-inducing feline retrovirus. Exp Haematol 17: 138–144

Dow SW, Hoover EA (1992) Neurologic disease associated with feline retroviral infection. In: Kirk RW, Bonagura J (eds) Current veterinary therapy XI. Saunders, Philadelphia, pp 1010–1012

Duvall E, Wyllie AH (1986) Death and the cell. Immunol Today 7: 115–119

Elder JM, Mullins JI (1983) Nucleotide sequence of the envelop gene of Gardner-Arnstein feline leukemia virus B reveals unique sequences of the envelop genes of two isolates of feline leukemia virus subgroup B. J Virol 46: 871–888

Essex ME, Cotter SM, Hardy WD Jr et al. (1975). Feline oncornavirus-associated cell membrane antigen. IV. Antibody titers in cats with naturally occurring leukemia, lymphoma and other diseases. J Natl Cancer Inst 55: 463–467

Estaquier J, Idziorek T, de Bels F, Barre-Sinoussi F, Hurtrel B, Aubertin AM, Venet A, Mehtail M, Muchmore E, Michel P (1994) Programmed cell death and AIDS: significance of T-cell apoptosis in pathogenic and nonpathogenic primate lentiviral infections. Proc Natl Acad Sci USA 91(20): 9431–5

Fisher AG, Ratner L, Mitsuya H, Marselle LM et al. (1986) Infectious mutants of HTLV-III with deletions in the 3' region and markedly reduced cytopathic effects. Science 233: 655–659

Forrest D, Onions D, Lees G, Neil JC (1987) Altered structure and expression of c-myc in feline T-cell tumors. Virology 158: 194–205

Francis DP, Essex M (1978) Leukemia lymphoma: infrequent manifestations of common viral infections? A review. J infect Dis 138: 916–923

Francis DP, Essex M (1980) Epidemiology of feline leukemia. Dev Cancer Res 4: 127–131

Francis DP, Essex M, Hardy WD Jr (1977) Excretion of feline leukaemia virus by naturally infected pet cats. Nature 269: 252–254

Francis DP, Essex M, Cotter S et al. (1979) Feline leukemia virus infections: The significance of chronic viremia. Leuk Res 3: 435–441

Fulton R, Forrest D, McFarlane R, Onions DE, Neil JC (1987) Retroviral transduction of T-cell antigen receptor beta-chain and myc genes. Nature 326: 190–194

Fulton R, Plumb M, Shield L, Neil JC (1990) Structural diversity and nuclear protein binding sites in the long terminal repeat of feline leukemia virus. J Virol 64: 1675–1682

Fultz PN, Zack P (1994) Unique lentivirus-host interactions: SIVsmmPBj infection of macaques. Virus Res 32: 205–225

Gardner MB (1985) Retroviral spongiform polioencephalomyelopathy. Rev Infect Dis 7: 99–110

Gardner MB, Henderson BE, Officer JE et al. (1973) A spontaneous lower motor neuron disease apparently caused by indigenous type C RNA in wild mice. J Natl Cancer Inst 51: 1243–1254

Goryo M, Hayashi S, Yoshizawa K, Umemura T, Itakura C, Yamashiro (1989) Ultrastructure of the thymus in chicks inoculated with chicken anaemia agent (MSB1-TK5803 strain). Avian Pathol 18: 605–617

Gougeon ML, Garcia S, Heeney J, Tschopp R, Lecoeur H, Guetard D, Rama V, Dauguet C, Montagnier L (1993) Programmed cell death in AIDS-related HIV and SIV infections. AIDS Res Hum Retroviruses 9(6): 553–63

Grant CK, Ernisse BJ, Pontefract R (1984) Comparison of feline leukemia virus-infected and normal cat T-cell lines in interleukin 2-conditioned medium. Cancer Res 44: 498–502

Groux H, Torpier G, Monte D, Mouton Y, Capron A, Ameisen JC (1992) Activation-induced death by apoptosis in CD4+ T cells from human immunodeficiency virus-infected asymptomatic individuals. J Exp Med 175: 331–340

Haase AT (1986) Pathogenesis of lentivirus infection. Nature 322: 130–136

Haffer KN, Sharpee RL, Beckenhauer W et al. (1987) Is the feline leukemia virus responsible for neurologic abnormalities in cats? Vet Med Aug: 802–805

Hamilton KL (1984) Feline leukemia virus receptors: enumeration Master's Thesis, Ohio State Univ, Columbus, Ohio and implications for pathogenesis.

Hardy WD Jr (1981) Feline leukemia virus nonneoplastic disorders. J Am Anim Hosp Assoc 17: 941–949

Hardy WD Jr, Hirshaut Y, Hess P (1973a) Detection of the feline leukemia virus and other mammalian oncornaviruses by immunofluorescence. In: Dutcher RM, Chieco-Bianchi L (eds) Unifying concepts of leukemia. Karger, Basel, pp 778–799

Hardy WD Jr, Old LJ, Hess PW et al. (1973b) Horizontal transmission of the feline leukaemia virus. Nature 244: 266–269

Hartke JR (1992) Immunopathogenesis of feline leukemia virus infection. PhD Dissertation, Ohio State Univ, Columbus, Ohio

Hildreth JE, Orentas RJ (1989) Involvement of a leukocyte adhesion receptor (LFA-1) in HIV-induced syncytium formation. Science 244: 1075–1078

Hoover EA, McCullough CB, Griesemer RA (1972) Intranasal transmission of feline leukemia. J Natl Cancer Inst 48: 973–983

Hoover EA, Perryman LE, Kociba GJ (1973) Early lesions in cats inoculated with feline leukemia virus. Cancer Res 33: 145–152

Hoover EA, Kociba GJ, Hardy WD Jr, Yohn DS (1974) Erythroid hypoplasia in cats inoculated with feline leukemia virus. J Natl Cancer Inst 53: 1271–1276

Hoover EA, Rojko JL, Wilson PL, Olsen RG (1980) Macrophages and the susceptibility of cats to feline leukemia virus infection. Dev Cancer Res 4: 182–186

Hoover EA, Rojko JL, Wilson PL, Olsen RG (1981) Determinants of susceptibility *versus* resistance to feline leukemia virus infection. I. Role of macrophages. J Natl Cancer Inst 67: 889–898

Hoover EA, Rojok JL, Quackenbush SL (1983) Congenital feline leukemia virus infection. Leuk Rev Int 1: 7–8

Hoover EA, Mullins JI, Quackenbush SL, Gasper PW (1987) Experimental transmission and pathogenesis of immunodeficiency syndrome in cats. Blood 70: 1880–1892

Hugin AW, Vacchio MS, Morse HC III (1991) A virus-encoded "superantigen" in a retrovirus-induced immunodeficiency syndrome of mice. Science 252: 424–427

Imberti L, Sottin l, Bettinardi A, Puoti M, Primi D (1991) Selective depletion in HIV infection of T cells that bear specific T cell receptor V-*beta* sequences. Science 245: 860–862

Janeway C Jr (1990) Self superantigens. Cell 63: 659–661

Jarrett O, Russell PH (1978) Differential growth + transmission in cats of feline leukemia viruses of subgroups A and B. Int J Cancer 21: 466–472

Jarrett WFH, Jarrett O, Mackey L, Larid H, Hardy W Jr, Essex M (1973) Horizontal transmission of leukemia virus and leukemia in the cat. J Natl Cancer Inst 51: 833–841

Jarrett O, Hardy WD Jr, Golder MC, Hay D (1978) The frequency of occurrence of feline leukemia virus subgroups in cats. Int J Cancer 21: 334–337

Jarrett O, Golder MC, Toth S, Onions DE, Stewart MA (1984) Interaction between feline leukaemia virus subgroups in the pathogenesis of erythroid hypoplasia. Int J Cancer 34: 283–288

Jarrett WFH, Martin WB, Crighton AW, Dalton RG, Stewart MF (1964) Leukaemia in the cat. Transmission experiments with leukaemia (lymphosarcoma). Nature 202: 566

Jeurissen SHM, Wagenaar F, Pol JMA, Van Der Eb AJ, Noteborn MHM (1992) Chicken anemia virus causes apoptosis of thymocytes after in vivo infection and of cell lines after in vitro infection. J Virol 66: 7383–7388

Khan KNM, Kociba GJ, Wellman ML, Reiter JA (1992a) Cytotoxicity of FeLV subgroup C-infected fibroblasts is mediated by adherent bone marrow mononuclear cells. In vitro Cell Dev Biol 28A: 260–266

Khan KNM, Kociba GJ, Wellman ML et al. (1992b) Effects of tumor necrosis factor-*alpha* on normal feline hematopoietic progenitor cells. Exp Hematol 20: 900–903

Khan KNM, Kociba GJ, Wellman ML (1992c) Synergistic effects of tumor necrosis factor-*alpha* and interferon-*gamma* on feline hemopoietic progenitors. Comp Haematol Int 2: 79–83

Khan KNM, Kociba GJ, Wellman ML (1993) Macrophage tropism of feline leukemia virus (FeLV) of subgroup C and increased production of tumor necrosis factor-α by FeLV-infected macrophages. Blood 81: 2585–2592

Kisielow P, Bluthman H, Staerz UD, Steinmetz M, Von Boehmer H (1988) Tolerance in T-cell receptor transgenic mice involves deletion of nonmature CD4+8+thymocytes. Nature 333: 742–746

Kociba GJ (1986) Hematologic consequences of feline leukemia virus infection. In: Kirk RW (ed) Current veterinary therapy, vol. XIII. Saunders, Philadelphia, pp 488–490

Kociba GJ, Halper JW (1987) Demonstration of retroviral proteins associated with erythroid progenitors of cats with feline leukemia virus-induced erythroid aplasia. Leuk Res 11: 1135–1140

Kociba GJ, Hoover EA, Sciulli VM, Olsen RG (1982) Feline leukemia virus-induced erythroid aplasia: Corticosteroid enhancement in vivo and FeLV suppression of erythroid colony development in vitro. In: Yohn DS, Blakeslee JR (eds) Advances in comparative leukemia research. Elsevier, New York, pp 211–212

Kociba GJ, Lange RD, Dunn CD, Hoover EA (1983) Serum erythropoietin changes in cats with feline leukemia virus-induced erythroid aplasia. Vet Pathol 20: 548–552

Kristal BS, Reinhart TA, Hoover EA, Mullins JI (1993) Interference with superinfection and with cell killing and determination of host range and growth kinetics mediated by feline leukemia virus surface glycoproteins. J Virol 67: 4142–4153

Landshulz WH, Johnson PF, Adashi EY, Graves BJ, McKnight SL (1988) Isolation of a recombinant copy gene encoding C/EBP. Gen Dev 2: 786–800

Laurent-Crawford AG, Krust B, Mulle S, Riviere Y, Rey-Cuille M-A, Bechet J-M, Montagnier L, Hovanessian AG (1991) The cytopathic effect of HIV is associated with apoptosis. Virology 185: 829–839

Levesque KS, Bonham L, Levy LS (1990) *flvi-1*, a common integration domain of feline leukemia virus in naturally occurring lymphomas of a particular type. J Virol 64: 3455–3462

Levy LS, Gardner MB, Casey JW (1984) Isolation of a feline leukaemia provirus containing the oncogene *myc* from a feline lymphosarcoma. Nature 308: 853–856

Lewis MG, Fertel RH, Olsen RG (1985) Reversal of feline retroviral suppression by indomethacin. Leuk Res 9: 1451–1456

Liesveld JL, Abboud CN, Duerst RE, Ryan DH, Brennan JK, Lictman MA (1989) Characterization of human bone marrow stromal cells. Role in progenitor cell binding and granulopoiesis. Blood 73: 1794–1800

Linenberger ML, Abkowitz JL (1992a) In vivo infection of marrow stromal fibroblasts by feline leukemia virus. Exp Haematol 20: 1022–1027

Linenberger ML, Abkowits JL (1992b) Studies in feline long-term marrow culture: hematopoiesis in normal and feline leukemia virus-infected stromal cells. Blood 80: 651–652

Liu Y, Janeway C (1990) Interferon gamma plays a critical role in induced cell death of effector T cells: a possible third mechanism of self-tolerence. J Exp Med 172: 735–1739

Lynn WS, Tweedale A, Cloyd MW (1988) Human immunodeficiency virus (HIV-1) cytotoxicity: perturbation of cell membrane and depression of phospholipid synthesis. Virology 163: 43–51

Mackey LJ, Jarrett WFH, Jarrett O, Laird H (1975) Anemia associated with feline leukemia virus infection in cats. J Natl Cancer Inst 54: 209–271

Mathes LE, Olsen RG, Hebebrand LC, Hoover EA, Schaller JP (1978) Abrogation of lymphocyte blastogenesis by a feline leukaemia virus protein. Nature 274: 687–689

McCabe MJ Jr, Orrenius S (1992) Deletion and depletion: the involvement of viruses and environmental factors in T-lymphocyte apoptosis. Lab Invest 66: 403–406

McClelland AJ, Hardy WD Jr, Zuckerman EE (1980) Prognosis of healthy feline leukemia virus infected cats. Dev Cancer Res 4: 121–126

McConkey DJ, Orrenius S, Jondal M (1990) Cellular signalling in programmed cell death (apoptosis). Immunol Today 11: 120–121

McDougall AS, Terry A, Tzavaras T, Cheney C, Rojko JL, Neil JC (1994) Defective endogenous proviruses are expressed in feline lymphoid cells: evidence for a role in natural resistance to subgroup B feline leukemia viruses. J Virol 68: 2151–2160

Meyaard L, Otto SA, Jonker RR, Mijnster MJ, Keet RP, Meidema F (1992) Programmed cell death of T cells in HIV-1 infection. Science 257: 217–219

Miller MA, Montelaro RC (1992) Amphipathic helical segments of human immunodeficiency virus type 1 transmembrane proteins and their potential role in viral cytopathicity. In: Aloia RC (ed) Advances in membrane fluidity, Vol 6. Liss, New York, pp 351–364

Miller MA, Garry RF, Jaynes JM, Montelaro RC (1991) A structural correlation between lentivirus transmembrane proteins and natural cytolytic peptides. AIDS Res Hum Retrovir 7: 511–519

Miller MA, Cloyd MW, Liebmann J, Rinaldo CR Jr, Islam KR, Wang SZS, Mietzner TA, Montelaro RC (1993a) Alterations in cell membrane permeability by the lentivirus lytic peptide (LLP-1) of HIV-1 transmembrane protein. Virology 196: 89–100

Miller MA, Mietzner TA, Cloyd MW, Robey WG, Montelaro RC (1993b) Identification of a calmodulin binding and inhibitory peptide domain in the HIV-1 transmembrane glycoprotein. AIDS Res Hum Retrovir 9: 1057–1066

Muller J, Krenn V, Schindler C, Czub S, Stahl-Henning C, Coulibaly C, Hunsmann G, Kneitz G, Kerkau T, Rethwilm A, terMeulen V, Muller-Hermelink HK (1993) Alterations of thymus cortical epithelium and interdigitating dendritic cells but no increase of thymocyte cell death in the early course of simian immunodeficiency virus infection. Amer J Pathol 143(3): 699–713

Mullins JI, Brody DS, Binari RC Jr, Cotter SM (1984) Viral transduction of c-*myc* gene in naturally occurring feline leukaemias. Nature 308: 856–858

Mullins JI, Chen CS, Hoover EA (1986) Disease-specific and tissue-specific production of unintegrated feline leukaemia virus variant DNA in feline AIDS. Nature 319: 332–336

Mullins JI, Hoover EA, Overbaugh J, Quackenbush SL, Donahue PR, Poss ML (1989) FeLV-FAIDS-induced immunodeficiency syndrome in cats. Vet Immunol Immunopathol 21: 25–37

Mullins JI, Hoover EA, Quackenbush SL (1991) Disease progression and viral genome variants in experimental feline leukemia virus-induced immunodeficiency syndrome. J Acquir Immune Def Syndr 4: 547–57

Neil JC, Hughes D, McFarlane R, Onions DE, Lees G, Jarrett O (1984) Transduction and rearrangement of the *myc* gene in naturally occurring feline leukaemias. Nature 308: 814–820

Nunberg JH, Williams ME, Innis MA (1984) Nucleotide sequences of the envelope genes of two isolates of feline leukemia virus subgroup B. J Virol 49: 629–632

Ohashi TG, Watari T, Tsujimoto H, Hasegawa A (1992) Elevation of feline interleukin-6-like activity in feline immunodeficiency virus infection. Clin Immunol Immunopathol 65(3): 207–211

Ohno K, Nakano T, Matsumoto Y, Watari T, Goitsuka R, Nakayama H, Tsujimoto H, Hasegawa A (1993) Apoptosis induced by tumor necrosis factor in cells chronically infected with feline immunodeficiency virus. J Virol 67(5): 2429–33

Ohno K, Okamoto Y, Miyazawa T, Mikami T, Watari T, Goitsuka R, Tsujimoto H, Hasegawa A (1994) Induction of apoptosis in a T lymphoblastoid cell line infected with feline immunodeficiency virus. Arch Virol 135(1–2): 153–8

Onions DE, Jarrett O, Testa NG, Frassoni F, Toth S (1982) Selective effect of feline leukaemia virus on early erythroid precursors. Nature 296: 156–159

Orosz CG, Zinn NE, Olsen RG, Mathes LE (1985) Retrovirus-mediated immunosuppression. I. FeLV-UV and specific FeLV proteins after T-lymphocyte behavior by inducing hyporesponsiveness to lymphokines. J Immunol 134: 3396–3403

Overbaugh J, Donahue PR, Hoover EA, Quackenbush SL, Mullins JI (1988) Molecular cloning of a feline leukemia virus that induces fatal immunodeficiency disease in cats. Science 239: 906–910

Overbaugh J, Hoover EA, Mullins JI, Burns DPW, Rudensey L, Quackenbush SL, Stallard V, Donahue PR (1992) Structure and pathogenicity of individual variants within an immunodeficiency disease-inducing isolate of FeLV. Virology 188: 558–569

Pandey R, Ghosh AK, Kumar DV, Bachman BA, Shibata D, Roy-Burman P (1991) Recombination between FeLV subgroup B or C and endogenous envelope elements alters the in vitro biological activities of the virus. J Virol 65: 6495–6508

Papiernik M, Brossard Y, Mulliez N, Roume J, Brechot C, Barin F, Goudeau A, Bach JF, Griscelli C, Henrion R (1992) Thymic abnormalities in fetuses aborted from human immunodeficiency virus type 1 seropositive women. Pediatrics 89(2): 297–301

Petry H, Dittmer U, Stahl-Henning C, Coulibaly C, Makoschey B, Fuchs D, Wachtert H, Tolle T, Morys-Wortmann C, Kaup FJ (1995) Reactivation of human immunodeficiency virus type 2 in macaques after simian immunodeficiency virus SIVmac superinfection. J Virol 69(3): 1564–1574

Poss ML, Mullins JI, Hoover EA (1989) Posttranslational modifications distinguish the envelope glycoprotein of the immunodeficiency disease-inducing feline leukemia virus retrovirus. J Virol 63: 189–195

Poss ML, Quackenbush SL, Mullins JI, Hoover EA (1990) Characterization and significance of delayed processing of the feline leukemia virus FeLV-FAIDS envelope glycoprotein. J Virol 64: 4338–4345

Price RW, Brew B, Sidtis J et al. (1988) The brain in AIDS: central nervous system HIV-1 infection and AIDS dementia complex. Science 239: 586–592

Quackenbush SL, Mullins JI, Hoover EA (1989) Colony-forming T lymphocyte deficits in the development of feline retrovirus-induced immunodeficiency syndrome. Blood 73: 509–516

Quackenbush SL, Donahue PR, Dean GA, Nyles MH, Ackley CD, Cooper MD, Mullins JI, Hoover EA (1990) Lymphocyte subset alterations and viral determinants of immunodeficiency disease induction by the feline leukemia virus FeLV-FAIDS. J Virol 64: 5465–5474

Reinacher M (1987) Feline leukemia virus-associated enteritis – condition with features of feline panleukopenia. Vet Pathol 24: 1–4

Riedel N, Hoover EA, Gasper PW et al. (1986) Molecular analysis and pathogenesis of the feline aplastic anemia retrovirus, feline leukemia virus C-Sarma. J Virol 60: 242–250

Riedel N, Hoover EA, Dornsife RE, Mullins JI (1988) Pathogenic and host range determinants of the feline aplastic anemia retrovirus. Proc Natl Acad Sci USA 85: 2758–2762

Rigby MA, Rojko JL, Stewart MA, Kociba GJ, Cheney CM, Rezanka LJ, Mathes LE, Hartke JR, Jarrett O, Neil JC (1992) Partial dissociation of subgroup C phenotype and in vivo behaviour in feline leukaemia viruses with chimaeric envelope genes. J Gen Virol 73: 2839–2847

Ringler DJ, Hunt RD, Desrosiers RC, Daniel MD, Chalifaux LV, King NW (1988) Simian immunodeficiency virus-induced meningoencephalomyelitis: natural history and retrospective study. Ann Neurol (Suppl) 23: S101–107

Rojko JL, Hardy WD Jr (1994) the feline leukemia virus and other retroviruses. In: Sherding RJ (ed) The cat: diseases and clinical management. Churchill Livingstone, New York, pp 263–432

Rojko JL, Hoover EA, Mathes LE, Hause WR, Schaller JP, Olsen RG (1978) Detection of feline leukemia virus in tissues of cats with a paraffin-embedding immunofluorescence procedure. J Natl Cancer Inst 63: 1315–1321

Rojko JL, Hoover EA, Mathes LE, Schaller JP, Olsen RG (1979) Pathogenesis of experimental feline leukemia virus infection. J Natl Cancer Inst 63: 759–768

Rojko JL, Hoover, EA, Finn BL, Olsen RG (1981) Determinants of susceptibility versus resistance to feline leukemia virus infection. II. Susceptibility of feline lymphocytes to productive feline leukemia virus infection. J Natl Cancer Inst 67: 899–910

Rojko JL, Cheney CM, Gasper PW, Hoover EA, Mathes LE, Kociba GJ (1986) Infectious feline leukemia virus is erythrosuppressive in vitro. Leuk Res 10: 1193–1199

Rojko JL, Fulton RM, Rezanka LJ, Williams LL, Copelan E, Cheney CM, Reichel GS, Neil JC, Mathes LE, Fisher TG, Cloyd MW (1992) Lymphocytotoxic strains of feline leukemia virus induce apoptosis in feline T4-thymic lymphoma cells. Lab Invest 66: 418–426

Rosenberg YJ, White BD, Papermaster SF, Zack P, Jarling PB, Eddy GA, Burke DS, Lewis MG (1991) Variation in T lymphocyte activation and susceptibility to SIV-PBJ-14-induced acute death in macaques. J Med Primatol 20: 206–210

Rosenzweig M, Clark KP, Gaulton GN (1993) Selective thymocyte depletion in neonatal HIV-1 thymic infection. AIDS 7: 1601–1605

Ruegg CL, Strand M (1990) Inhibition of protein kinase C and anti-CD3-induced Ca^{2+} influx in Jurkat T cells by a synthetic peptide with sequence identity to HIV gp41. J Immunol 144: 3928–3925

Ruprecht R, Fratazzi C, Sharma PL, Greene MF, Penninck D, Wyand M (1993) Animal models for perinatal transmission of pathogenic viruses. Ann NY Acad Sci 693: 213–228

Sarma PS, Log T (1973) Subgroup classification of feline leukemia and sarcoma viruses by viral interference and neutralization tests. Virology 54: 160–169

Sarma PS, Log T, Skuntz S, Krishnan S, Burkley K (1978) Experimental horizontal transmission of feline leukemia viruses of subgroups A, B, and C. J Natl Cancer Inst 60: 871–874

Schnittman S, Denning SM, Greenhouse JJ, Justement JS, Baseler M, Kurtzberg J, Haynes BF, Fauci AS (1990) Evidence for susceptibility of intrathymic T-cell precursors and their progeny carrying T-cell antigen receptor phenotypes to human immunodeficiency virus infection: a mechanism for CD4+ (T4) lymphocyte depletion. Proc Natl Acad Sci USA 87: 7727–7731

Schnittman S, Singer K, Greenhouse J, Stanley SK, Whichard LP, Le PT, Haynes BF, Fauci AS (1991) Thymic microenvironment induces HIV expression: Physiologic secretion of IL-6 by thymic epithelial cells upregulates virus expression in chronically infected cells. J Immunol 147(8): 2553–2558

Sellins KS, Cohen JJ (1987) Gene induction by gamma-irradiation leads to DNA fragmentation in lymphocytes. J Immunol 139: 3199–3206

Shaw GM, Hahn BN, Arya SK, Groopman JE, Gallo RC, Wong-Staal F (1984) Molecular characterization of HTLV-III in the acquired immunodeficiency syndrome. Science 226: 1165–1171

Smith CA, Williams GT, Kingston R, Jenkinson EJ, Owen JJT (1989) Antibodies to CD3/T-cell rereptor complex induce death by apoptosis in immature T-cells in thymic culture. Nature 337: 181–184

Soe LH, Devi BG, Mullins JI, Roy-Burman P (1983) Molecular cloning and characterization of endogenous feline leukemia virus sequences from a cat genomic library. J Virol 46: 829–840

Soe LH, Shimizu RW, Landolph JR, Roy-Burman P (1985) Molecular analysis of several classes of endogenous feline leukemia virus elements. J Virol 56: 701–710

Steinberg HN, Crumpacker CS, Chatis PA (1991) In vitro suppression of normal human bone marrow progenitor cells by human immunodeficiency virus. J Virol 65: 1765–1769

Stewart MA, Warnock M, Wheeler A, Wilkie N, Mullins JI, Onions DE, Neil JC (1986) Nucleotide sequences of a feline leukemia virus subgroup A envelope gene and long terminal repeat and evidence for the recombinational origin of subgroup B viruses. J Virol 58: 825–834

Terai C, Kornbluth RS, Pauza CD, Richman DD, Carson DA (1991) Apoptosis as a mechanism of cell death in cultured T-lymphoblasts acutely infected with HIV-1. J Clin Invest 87: 1710–1715

Testa NG, Onions D, Jarrett O, Frassoni F, Eliason J (1983) Haematopoietic colony formation (BFU-E, GM-CFC) during the development of pure red cell hypoplasia induced in the cat by feline leukemia virus. Leuk Res 7: 103–116

Tochikura TS, Hayes KA, Cheney CM, Tanabe-Tochikura A, Rojko JL, Mathes LE, Olsen RG (1990) In vitro replication and cytopathogenicity of the feline immunodeficiency virus (FIV) for feline T4 thymic lymphoma 3201 cells. Virology 179: 492–498

Tomei LD, Kanter P, Wenner CD (1988) Inhibition of radiation-induced apoptosis in vitro by tumor promoters. Biochem Biophys Res Commun 155: 324–331

Tsai S, Patel V, Beaumont E, Lodish HF, Nathan DG, Sieff CA (1987) Differential binding of erythroid and myeloid progenitors to fibroblasts and fibronectin. Blood 69: 1581–1594

Tzavaras T, Stewart M, McDougall A, Fulton R, Testa N, Onions DE, Niel JC (1990) Molecular cloning and characterization of a defective recombinant feline leukaemia virus associated with myeloid leukaemia. J Gen Virol 71: 343–354

Ucker DS (1987) Cytotoxic T-lymphocytes and glucocorticoids activate an endogenous suicide process in target cells. Nature 327: 62–64

Valentin H, Nugeyre M-T, Vuillier F, Boumsell L, Schmid M, Barre-Sinoussi F, Pereira RA (1994) Two subpopulations of human triple-negative thymic cells are susceptible to infection by human immunodeficiency virus type 1 in vitro. J Virol 68(5): 3041–3050

Wardrop KJ, Kramer JW, Abkowitz JL, Clemons SG, Adamson JW (1986) Quantitative studies of erythropoietin in the clinically normal, phlebotomized, and feline leukemia virus-infected cat. Am J Vet Res 478: 2274–2277

Watanabe M, Ringler DJ, Nakamura M, DeLong PA, Letvin NL (1990) Simian immunodeficiency virus inhibits bone marrow hematopoietic progenitor cell growth. J Virol 64: 656–663

Weiser MG, Kociba GJ (1983) Erythrocyte macrocytosis in feline leukemia virus-associated anemia. Vet Pathol 20: 687–697

Weller SK, Joy AE, Temin HM (1980) Correlation between cell killing and massive second round superinfection by members of some subgroups of avian leukosis virus. J Virol 33: 494–506

Wellman ML, Kociba GJ (1988) Characterization of fibroblast colony-forming units in bone marrow from healthy cats. Am J Vet Res 49: 231–235

Wellman ML, Kociba GJ, Lewis MG, Mathes LE, Olsen RG (1984) Inhibition of erythroid colony-forming cells by a 15,000 dalton protein of feline leukemia virus. Cancer Res 44: 1527–1531

Wellman ML, Kociba GJ, Mathes LE, Olsen RG (1988) Suppression of feline bone marrow fibroblast colony-forming units by feline leukemia virus. Am J Vet Res 49: 227–231

Wellman ML, Kociba GJ, Mathes LE (1991) Variable suppression of feline bone marrow fibroblast colony-forming units by two isolates of feline leukemia virus. Am J Vet Res 52: 1924–1928

Wheeler D, Whalen LR, Gasper PW et al. (1990) A feline model for AIDS dementia complex (ADC)? 6th Int AIDS Conf, San Francisco, California

Williams GT, Smith CA, Spooncer E, Dexter TM, Taylor DR (1990) Haemopoietic colony stimulating factors promote cell survival by suppressing apoptosis. Nature 343: 76–79

Williams LL, Lewis MG, Olsen RG, Lafrado LJ, Horrocks LA, Rojko JL (1993) Chronic feline leukemia virus infection alters arachidonic acid proportions in vivo and in vitro. Proc Soc Exp Biol Med 202: 239–245

Wyllie AH (1987) Apoptosis: cell death in tissue regulation. J Pathol 153: 313

Yamada T, Ohyama H (1988) Radiation-induced interphase death of rat thymocytes is internally programmed (apoptosis). Int J Radiat Biol 53: 65–75

Zack P, Kociba GJ (1988) Effects of increasing cyclic AMP or calcium on feline erythroid progenitors in vitro: normal cells are stimulated while cells from retrovirus-infected cats are suppressed. Int J Cell Cloning 6: 192–208

Neurotoxicity in Rat Cortical Cells
Caused by N-Methyl-D-Aspartate (NMDA) and gp120
of HIV-1: Induction and Pharmacological Intervention

W.E.G. Müller[1], G. Pergande[2], H. Ushijima[3], C. Schleger[1],
M. Kelve[4], and S. Perovic[1]

Abstract

Incubation of highly enriched neurons from rat cerebral cortex with the human
immunodeficiency virus type 1 (HIV-1) coat protein gp120 for 18 h results in
fragmentation of DNA at internucleosomal linkers, a feature of apoptosis. We
report that neurons respond to exposure to gp120 with an increased release of
arachidonic acid via activation of phospholipase A_2. This process is not inhibited
by antagonists of the N-methyl-D-aspartate (NMDA) receptor channels. To
investigate the influence of arachidonic acid on the sensitivity of NMDA receptor
towards its aganist, low concentrations of NMDA were coadministered with
arachidonic acid. Under these conditions the NMDA-mediated cytotoxicity was
enhanced. We conclude that gp120 causes an activation of phospholipase A_2,
resulting in an increased release of arachidonic acid which in turn sensitizes the
NMDA receptor.

Two compounds were found to act cytoprotectively against the deleterious
effect caused by gp120 on neurons: Memantine [1-amino-3,5-dimethylada-
mantane] and Flupirtine [2-amino-3-ethoxycarbonylamino-6-(4-fluoro-benzyl-
amino)-pyridine maleate]. Both compounds have been found to display a potent
cytoprotective effect on neurons treated with the excitatory amino acid NMDA
or with the human immunodeficiency virus type 1 (HIV-1) coat protein gp120.
The NMDA antagonist Memantine, a drug currently used in the therapy of
spasticity and Parkinson's disease, prevented the effects of gp120 at micro-
molar concentrations. Flupirtine was previously found to be a centrally acting,
nonopiate analgesic agent which additionally possesses anticonvulsant and
muscle-relaxant activity at doses similar to those producing analgesia. The
cytoprotective effect of Flupirtine in vitro was significant (above 10 µM). Considering the fact that both Memantine and Flupirtine display almost no clinical
side effects, these drugs may prove useful both in preventing primary infection of

[1] Institut für Physiologische Chemie, Abteilung Angewandte Molekularbiologie, Universität,
Duesbergweg 6, 55099 Mainz, Germany
[2] ASTA Medica AG, Abteilung Medizin Deutschland, Weismüllerstr. 45, 60314 Frankfurt,
Germany
[3] Institute of Public Health; 6-1, Shirokanedai 4-chome, Minato-ku, Tokyo 108, Japan
[4] Institute of Chemical Physics and Biophysics, Akadeemia tee 23, EE0026 Tallinn, Estonia

brain cells with the HIV virus, as well as in treating the neurological disorders often associated with the immunodeficiency syndrome such as AIDS-related dementia.

1   Introduction

Apoptosis, also termed programmed cell death, occurs both under physiological and pathological conditions (Williams 1991). Recently, it has been described that the human immunodeficiency virus type 1 [HIV-1] causes apoptosis in human CD4$^+$ lymphocytes (Gougeon et al. 1991). HIV-1 is a retrovirus which integrates into the genome of CD4$^+$ lymphocytes (reviewed in Wong-Staal and Gallo 1985). Besides its transforming ability, this virus causes cytopathicity to most cells (reviewed in Wong-Staal and Gallo 1985). With respect to CD4$^+$ cells, the HIV-1 envelope glycoprotein gp120 induces apoptosis not only in the soluble form but also in the membrane-associated state (Laurent-Crawford et al. 1993). In 1992, we described that rat neurons undergo apoptosis in the presence of free gp120 in vitro (Müller et al. 1992). The molecular mechanisms underlying this unexpected finding are not yet fully understood.

It is known that the excitatory amino acids (EAA), glutamate and asparate, play vitally important metabolic, neurotrophic, and neurotransmitter roles, but display also neurotoxic (excitotoxic) potential (reviewed in Olney 1993). These EAA act on a family of neuronal receptors that mediate Glu/Asp excitotoxicity. The mammalian Glu receptor family is subdivided into the classes of (1) ionotropic receptors, that are coupled to an ion channel for signal transduction, and (2) metabotropic receptors, that are linked with phosphoinositide hydrolysis (Sladaczek et al. 1985). According to their sensitivities to specific agonists, the ionotropic Glu receptors are further subdivided into (1) the N-methyl-D-aspartate (NMDA) receptors, (2) the quisqualic acid/amino-3-hydroxy-5-methylisoxazole-4-propanic acid receptors and (3) the kainic acid receptors (Watkins and Evans 1981).

We discovered that the safe noncompetitive NMDA antagonist, 1-amino-3,5-dimethyladamantane (Memantine; Bormann 1989), already in clinical use to treat Parkinson's disease, protects rat cortical cells against apoptosis induced by HIV-1 gp120 (Müller et al. 1992). Recently, we found that also a member of the class of triaminopyridines, Flupirtine-maleate [2-amino-3-ethoxycarbonyl-amino-6-(4-fluorobenzylamino)-pyridine maleate], which is applied in clinics as a centrally acting, nonopiate analgesic agent with anticonvulsant and muscle-relaxant activity (Szelenyi et al. 1989) protects rat cortical neurons against NMDA and HIV-1 gp 120-induced apoptosis (Perovic et al. 1994).

2 Induction of Apoptosis in Rat Cortical Cells by HIV-1 gp120 or NMDA In Vitro

2.1 DNA Fragmentation and Cell Morphology After gp120 Treatment

As reported earlier, neurons undergo apoptosis in response to HIV-gp120 (Dreyer et al. 1990; Müller et al. 1992). If rat cortical cells are exposed for 15 min to gp120 in Mg^{2+}-free solution and subsequently put into the incubator in the original growth medium/serum for 12 h, the neurons die (Müller et al. 1992). It was found that at a concentration of 200 pM, gp120 strongly increased the extent of DNA fragmentation from 7% (controls) to 62%; simultaneously the percentage of viable cells decreased from 94 to 33% (Table 1). This effect could be partially abolished by coincubation with two lectins, one from *Narcissus pseudonarcissus* (Weiler et al. 1990) and another from *Gerardia savaglia* (Kljajic et al. 1987), which have previously been shown to bind to gp120 with high affinity (Müller et al. 1988).

The integrity of DNA of neuronal cells after exposure to gp120 was analyzed time-dependently applying agarose gel electrophoresis. Control cells show only very little DNA fragmentation (Fig. 1, lane a). During incubation with gp120, the DNA undergoes increasing fragmentation (lanes b–e). After an incubation period of 18 h, DNA degradation products described for apoptotic cells, which are multiples of ≈180 base pairs in size, are identified (lane e).

Microscopical analysis revealed that in the presence of HIV-gp120 the cells lost their typical neuronal appearance; the neurons have almost completely lost the neurites and the cell bodies appear to have shrunken (Fig. 2b). In contrast, the untreated cells which are distributed nearly homogeneously with a few or no aggregates show well-developed neurite bundles (Fig. 2a).

2.2 Induction of Apoptosis in Cortical Cell Cultures by HIV-1 Particles

In a further series of experiments, intact virus preparations were used for the induction of apoptosis; a concentration of 30 ng/ml of HIV-gp120 equivalent was applied and added to higly enriched neurons. Under these conditions, the viability of the cells dropped from 76% to 4% during the 18-h incubation period (Table 2); the percentage of DNA fragmentation increased from 13 to 68%. In the control

Table 1. Effect of gp120 on DNA fragmentation and cell toxicity in rat neuronal cultures. Cells were prepared from the brains of 18-day-old Wistar rat embryos. They were incubated (12 h) in the absence or presence of gp120 under conditions described (Müller et al. 1992). Where indicated, Memantine was added to the cultures 2 h prior to the addition of gp120

Addition of gp120 (pM)	Compound (µM)	DNA fragmentation (%)	Percentage of viable cells
0	0	7.2	93.6
200	0	62.4	32.7
0	Memantine 100	8.5	97.6
200	Memantine 10	39.2	68.8
200	Memantine 100	21.8	85.0

Fig. 1. Effects of HIV-1 gp120 on the integrity of DNA from neuronal cells. The cultures were incubated in the presence of 100 pM for 0 to 18 h (as indicated; *lanes a–e*). DNA was extracted and analyzed by agarose gel electrophoresis. Per lane, 0.3 μg DNA was applied. The molecular masses are shown

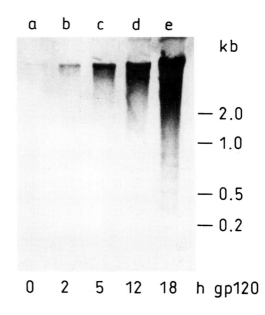

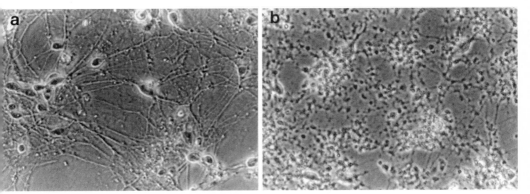

Fig. 2a, b. Induction of neuronal death in rat cortical cells, after exposure to HIV-gp120. Cultures remained untreated (**a**) or were treated with 100 pM of HIV-gp120 for 18 h (**b**). x 300

Table 2. Effect to HIV-gp120 equivalent (HIV-1 particles) on DNA fragmentation and cell toxicity. Highly enriched cortical cells were treated for 18 h in the absence or presence of HIV-gp120 equivalent as described (Perovic et al. 1994). Where indicated, Flupirtine was added to the cultures 2 h prior to the addition of the viral protein

Addition of HIV-gp120 equivalent (ng/ml)	Compound (μM)	DNA fragmentation (%)	Percentage of viable cells
0	0	12.8	76.4
30	0	68.4	4.4
0	Flupirtine 10	8.3	82.6
0	Flupirtine 40	6.2	96.7
30	Flupirtine 10	46.9	46.2
30	Flupirtine 20	28.4	59.7
30	Flupirtine 40	24.7	68.4

samples treated with a preparation of HIV-gp120, which had been preincubated with antibodies against gp120, no significant reduction in cell viability was measured (not shown).

2.3 Increased Release of Arachidonic Acid from Neurons After Incubation with gp120

Neuronal cell cultures were incubated in the presence of [³H] arachidonic acid for 20 h. During this period, ≈70% of the radioactivity added to the cultures was incorporated into the cells. Subsequently, they were incubated with different concentrations of gp120 for 0–15 min. Then the medium was removed, the cells centrifuged, and the supernatant was measured by liquid scintillation counting.

As summarized in Fig. 3, gp120 induced a stimulation of the release of arachidonic acid during a 20-h incubation period. The enhanced release of arachidonic acid was found to be dose-dependent; at a concentration of 50 pM of gp120 the release is stimulated to 122% [with respect to the controls (without gp120)] and at 100 pM to 170% (Fig. 3). At higher doses of gp120, the level increases further and reaches ≈180% with respect to the control (basal) value.

2.4 Inhibition of Arachidonic Acid Release by Phospholipase A₂ Inhibitor

A series of inhibitors have been applied to elucidate the mode of action of gp120-induced stimulation of arachidonic acid release from neuronal cells. While indomethacin (cyclooxygenase inhibitor; at a concentration of 50 µM), nordihydroguaiaretic acid (NDGA; lipoxygenase inhibitor; 50 µM) and H7 (kinase C inhibitor; 100 µM) failed to significantly modify the gp120-induced arachidonic

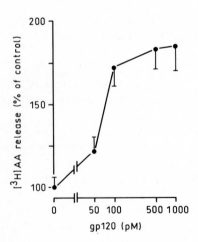

Fig. 3. Effect of gp120 on the release of [³H] arachidonic acid (AA) from neuronal cells. Cells were prelabeled with this compound and subsequently incubated with different concentrations of gp120 for 15 min and the radioactivity in the medium was determined. The values are given in percent of the values determined in the controls

acid release, mepacrine (50 µM), a phospholipase A_2 inhibitor, blocked the response to gp120 significantly and almost completely (Table 3).

2.5 Induction of Apoptosis in Cortical Cell Cultures of NMDA

NMDA displayed a neurotoxic effect; at a concentration of 100 µM (or 500 µM) of NMDA, the DNA fragmentation increased from 7% (controls) to 35% (68%). Simultaneously, the viability of the cells decreased from 74% (controls) to 43% (28%) (Table 4). At higher concentrations (3 mM) of NMDA the amount of living cells decreased further. After the 18-h incubation period, only 8% of the cells remained alive; in parallel, the degree of DNA fragmentation increased to 95% (Table 5). After electrophoresis in agarose gel a ladder-like pattern of DNA with degradation products in multiples of ≈180 base pairs was detected (Perovic et al. 1994). In the controls (no agonist present), no fragmentation was observed.

Untreated control cells are distributed in a nearly homogeneous fashion in the culture dish and show long neurites. During incubation in the presence of NMDA, the cells change their morphology; they lose their neurites, and their cell bodies shrink (Perovic et al. 1994).

2.6 Influence of NMDA Antagonists on Arachidonic Acid Release

In a previous study, we could demonstrate that the deleterious effect of gp120 leading to cell death in neurons can be prevented by Memantine, MK-801 (Müller et al. 1992), and Flupirtine (Perovic et al. 1994), three drugs that are potent blockers of N-methyl-D-aspartate (NMDA) receptor channels (Huettner and Bean 1988; Bormann 1989; Kornhuber et al. 1989). Therefore, we used these three compounds

Table 3. Effect of inhibitors on gp120-induced augmentation of [³H]arachidonic acid (AA) release from neurons. Cells were preincubated with [³H]arachidonic acid for 20 h and subsequently incubated for 15 min in the absence or presence of 100 pM gp120. Radioactivity in the medium was determined. The percentage of total arachidonic acid release in the controls (not treated with gp120) was 1.27%, while after incubation with the viral protein the release increased to 2.32%. The difference between treated and nontreated cultures (1.05%) was set to 100% (% of gp120 induced arachidonic acid release). The following inhibitors were used: Mepacrine (MEP), indomethacin (IND), NDGA, and H7

Addition of gp120 [pM]	Inhibitor (µM)	[³H]AA release % of gp120 induced
100	0	–
Net gp120-mediated release:		100
100	MEP 50	7
100	IND 50	99
100	NDGA 50	97
100	H7 100	93

Table 4. Effect of arachidonic acid on NMDA-induced DNA fragmentation and cell toxicity. DNA fragmentation was determined by the sedimentation technique (Burton 1956), and viability was assayed with fluorescein diacetate. (Hahn et al. 1988)

NMDA (μM)	Arachidonic acid (μM)	DNA fragmentation (%)	Percentage of viable cells
0	0	7.3	74.3
0	20	6.6	76.7
100	0	34.9	42.6
100	5	51.4	33.2
100	20	67.3	24.0
500	0	68.2	28.4
500	5	76.9	16.2
500	20	91.3	8.5

Table 5. Prevention of NMDA-induced DNA fragmentation and cell toxicity by Flupirtine. Rat neurons were incubated in the absence or presence of NMDA; Flupirtine was added to the cultures 2 h prior to the addition of NMDA. The incubation time was 18 h. DNA fragmentation was determined by the sedimentation technique and viability was assayed with fluorescein diacetate

Addition of NMDA (mM)	Flupirtine (μM)	DNA fragmentation (%)	Viable cells (%)
0	0	12.8	76.4
0	1	11.0	79.4
0	10	8.3	82.6
0	20	6.9	92.4
0	40	6.2	96.7
3	0	94.7	7.8
3	1	63.5	29.1
3	10	16.0	78.5
3	20	9.9	86.8
3	40	5.2	92.0

to elucidate if the observed enhanced release of incorporated arachidonic acid from neuronal cells is induced by gp120 through an activation of the NMDA receptor. The cells were preincubated with the compounds and subsequently treated with gp120. Neither MK-801 nor Memantine or Flupirtine significantly affected the basal level of arachidonic acid release in the presence of gp120 (Table 6). Again, it was measured that incubation of the cells with 100 pM of gp120 increased the release of arachidonic acid.

In addition, the potential effect of the three NMDA antagonists to function cytoprotectively was measured. It was found that all compounds protect the cells against apoptotic death dose-dependently. In the presence of 1 μM MK-801, the percentage of viable cells which were treated with gp120 increased from 27 to 72%. Likewise significant was the cytoprotective effect of Memantine and Flupirtine; at a concentration of 5 μM Memantine or 10 μM Flupirtine, the number of viable cells increased to 56 and 53%, respectively (Table 6). At higher

Table 6. Effect of the NMDA antagonists MK-801, Memantine, and Flupirtine on gp120-induced stimulation of [³H] arachidonic acid (AA) release from neuronal cells and on cell viability. The cells were pretreated with the compounds for 2 h and subsequently treated with 100 pM of gp120

Pretreatment (µM)		Incubation	Total AA release (%)	Viable cells (%)
None		–	1.24	84.6
None		gp120	2.36	26.9
MK-801:	1	gp120	2.28	72.3
	10	gp120	2.29	76.8
Memantine:	5	gp120	2.33	56.0
	50	gp120	2.25	79.3
Flupirtine	10	gp120	2.30	53.2
	30	gp120	2.42	77.4

concentrations, the cytoprotective effect of MK-801, Memantine, and Flupirtine was more pronounced.

2.7 Arachidonic Acid Augments the NMDA-Caused DNA Fragmentation

The above data indicate that the increased release of arachidonic acid in response to the interaction of the neurons with gp120 is not a result of an activation of the NMDA receptor. Moreover, the data show that NMDA antagonists prevent gp120-mediated cytotoxicity. Therefore, the possibility had to be tested if arachidonic acid sensitizes the NMDA receptor and allows the glutamate which is present in the culture medium (Lipton et al. 1991; Ushijima et al. 1993) to activate the receptor.

As summarized in Table 4, arachidonic acid added alone did not result in a significant alteration of either the extent of DNA fragmentation or cell viability. However, if arachidonic acid was added together with NMDA, the toxic effect on neurons caused by the latter compound increased. For example, a coincubation of 100 µM of NMDA together with 5 µM of arachidonic acid resulted in an increased DNA fragmentation (51%) as well as a decrease in the number of viable cells (33%; Table 4).

3 Prevention of Apoptosis in Cortical Cells In Vitro by Memantine and Flupirtine

3.1 Memantine

Memantine [1-amino-3,5-dimethyladamantane] was described as a potent blocker of NMDA receptor channels (Huettner and Bean 1988; Bormann 1989; Kornhuber et al. 1989).

Previously, we reported that gp120-induced apoptosis of rat cortical cells is strongly reduced by Memantine (Müller et al. 1992). The data revealed that the process of DNA fragmentation in neurons in response to an incubation of the cultures with 200 pM of gp120 for 12 h is strongly inhibited by coincubation with 10 or 100 μM of Memantine (Table 1).

3.2 Flupirtine

Flupirtine-maleate [2-amino-3-ethoxycarbonylamino-6-(4-fluoro-benzylamino)-pyridine maleate], is a member of the class of triaminopyridines. Recently, Schwarz et al. (1994) have found indications that the myorelaxant effect of Flupirtine might be mediated via an interaction with the NMDA receptor. This conclusion was supported by immunohistochemical data from Osborne et al. (1994), which revealed that Flupirtine nullified the NMDA-induced changes in the rabbit retina system.

To substantiate the assumption that Flupirtine activates NMDA receptors, in vitro studies with rat primary neuronal cell cultures were performed (Perovic et al. 1994).

3.2.1 Prevention of NMDA or HIV-gp120-Induced Apoptosis in Cortical Cells by Flupirtine

Flupirtine has a strong cytoprotective effect on neurons against the toxic action displayed by NMDA or HIV-gp120. Incubation of cortical cells with Flupirtine results in a dose-dependent increase in the amount of viable cells in cultures treated with NMDA (Table 5). In the presence of 10 μM Flupirtine, viability increases from 8 to 79%; at 40 μM Flupirtine, the viability is 92%. Simultaneously with the increase in the cell viability in the presence of Flupirtine, the extent of DNA fragmentation decreases; at 10 μM Flupirtine the yield significantly decreases from 95% (in the absence of the drug) to 16%, and at 40 μM to 5%.

Flupirtine is likewise effective against the neurotoxic effect displayed by HIV-gp120 (Table 2). Addition of the drug at concentrations >10 μM caused a significant increase in cell viability; at this concentration the percentage of viable cells increases from 4% (absence of the compound) to 46%, and at 40 μM to 68%.

The cytoprotective effect of Flupirtine on NMDA- and HIV-gp120-induced cytotoxicity could also be documented by a strong reduction in DNA fragmentation, as visualized by agarose gel electrophoresis (Perovic et al. 1994). Fragmentation is almost completely prevented at 20 μM Flupirtine.

The morphology of the NMDA- and HIV-gp120-treated cells which have been preincubated with 20 μM Flupirtine was indistiguishable from that of the controls (Perovic et al. 1994). The neurons retain their typical appearance.

3.2.2 Cytoprotective Effect of Flupirtine on Untreated Rat Cortical Cells

During the culturing of untreated cells (controls) for 18 h, the percentage of living cells decreases significantly to 76% and the percentage of DNA fragmentation increases to 13% (Table 2). Incubation of the cells with different concentrations of Flupirtine results in a gradual increase of the percentage of cell viability; simultaneously, the percentage of DNA fragmentation decreases. At a concentration of 10 μM, the percentage of viable cells increases from 76% (controls) to 83%. At 40 μM Flupirtine, cell viability is 97% and the degree of DNA fragmentation drops from 13% (controls) to 6%.

4 Conclusion

4.1 Cell Biological Findings

Earlier studies by us revealed that gp120 induces apoptosis in neuronal cells (Müller et al. 1992). However, until now the molecular mechanism by which the viral protein causes this event has been poorly understood. Previous studies suggested that gp120 does not interact with the NMDA receptor directly (Savio and Levi 1993; Ushijima et al. 1993). Two, perhaps alternative, routes can be postulated; (1) the neurons – after contact with gp120 – release substances into the extracellular milieu which are either neurotoxic or display a sensitization of the NMDA receptor or (2) gp120 alters in neurons intracellularly signal transduction pathways leading to apoptosis.

The reported data indicate that rat neuronal cells respond to an incubation in the presence of gp120 with an increased release of arachidonic acid. This result supports previous studies which report that neurons liberate arachidonic acid from membrane phospholipids in response to NMDA treatment via phospholipase A_2 (Sanfeliu et al. 1990; Tapia-Arancibia et al. 1992). It appears to be indicative that the extent of the augmented arachidonic acid release from phospholipids is almost identical for both NMDA and gp120. Previously, it was reported that NMDA stimulates the release of arachidonic acid by two- to three-fold (Sanfeliu et al. 1990; Tapia-Arancibia et al. 1992), while gp120 induces the release close to two-fold (this chapter).

As a first approach to solving the mechanism by which arachidonic acid is released from phospholipids, inhibition studies were performed. Among the inhibitors tested, only the phospholipase A_2 inhibitor mepacrine (Billah et al. 1985) affected gp120-induced arachidonic acid release strongly. The other agents indomethacin (a cyclooxygenase inhibitor; Sung et al. 1988), NDGA (lipoxygenase inhibitor; Lynch et al. 1989) and H7 (protein kinase C inhibitor; Hidaka et al. 1984) displayed no effect. Based on the mepacrine inhibition data, we conclude that the effect of gp120 in this system is predominantly due to an activation of phospholipase A_2.

Earlier findings suggested that increased activity of phospholipase A_2 is mediated via a gp120-induced activation of the NMDA receptor. In order to test this possibility, studies with Flupirtine (Osborne et al. 1994; Schwarz 1994) and two non-competitive NMDA antagonists, MK-801 (Huettner and Bean 1988) and Memantine (Bormann 1989), were performed. However, these three inhibitors did not change the gp120-induced release of arachidonic acid from neurons. Simultaneously, the compounds displayed their cytoprotective effect as reported already earlier (Müller et al. 1992). Therefore, we conclude that gp120 does not interact directly with the NMDA receptor(s) (Ushijima et al. 1993). The molecular mechanism by which gp120 causes an activation of phospholipase A_2 is not known.

The coincubation experiments of arachidonic acid and NMDA, the agonist of the NMDA receptor, revealed that in the presence of both agents both neuronal cell death and DNA fragmentation are augmented compared to an administration of NMDA alone. Therefore we suggest that arachidonic acid sensitizes the NMDA receptor and in turn augments the NMDA-mediated cell toxicity. This conclusion is supported by an earlier study showing that arachidonic acid potentiates the NMDA receptor function in the presence of appropriate agonists (Miller et al. 1992).

Based on these data, we propose the following sequence of events, following the interaction of gp120 with neuronal cells (Fig. 4). Binding of gp120 to neurons – activation of phospholipase A_2 resulting in an increased release of arachidonic acid into the extracellular milieu – binding of arachidonic acid to the NMDA receptor – sensitization of the NMDA receptor resulting in a lowering of the threshold for the agonists NMDA and glutamate. Glutamate is present in the culture medium in low concentrations (≈ 20 µM) usually not high enough to activate this receptor; this concentration might, however, be sufficient to function as agonist after sensitization of the receptor, as suggested earlier (Lipton et al. 1991; Ushijima et al. 1993).

NMDA is known to induce the expression of the oncogene c-*fos* via activation of the NMDA receptor; this alteration can be blocked by phospholipase A_2 inhibitors (Lerea and McNamara 1993). This finding is interesting in view of data suggesting that c-*fos* expression is involved in apoptosis (Wu et al. 1993). Future studies should clarify the links between gp120 association to neurons, expression of c-*fos* and induction of apoptosis.

4.2 Pharmacological Interventions

Here, we describe that Memantine and Flupirtine display a protective effect against NMDA- and HIV-gp120-induced neuronal cell death. NMDA as well as gp120 (Dreyer et al. 1990; Ushijima et al. 1993) cause an increase in Ca^{2+} concentration in neurons, a process which ultimately results in an apoptotic death of the cells (Müller et al. 1992). It is not yet fully understood whether HIV-gp120 affects neurons directly or indirectly via other cell systems. Previous studies in both in vitro cultures and in vivo have suggested that HIV-gp120 interacts with macrophages, resulting in the production of neurotoxins (Giulian et al. 1990). Recent results with purified neuronal cells strongly indicate that the viral protein has a direct neurotoxic effect (Savio and Levi 1993).

Flupirtine is clinically safe; adverse reactions are minimal in incidence, nature, and degree, with drowsiness as the most frequently reported reaction

Fig. 4. Proposed series of events leading to a sensitization of the NMDA receptor which might be induced after interaction of gp120 with neurons. Ultimately, exogenous NMDA or glutamate activates the receptor which, in turn, induces apoptosis perhaps via an increased expression of *c-fos* oncogene

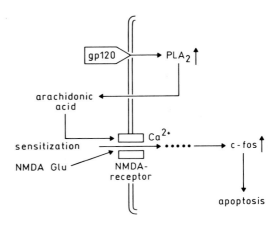

(approximately 10% of the patients; McMahon et al. 1987). Cell biological studies have not yet been published.

Pharmacokinetic data revealed that at the therapeutic daily dose of 600 mg administered orally to patients (Friedel and Fitton 1993), a plasma level of 2.5 µg/ml is obtained (Maier-Lenz and Thieme 1983). Liquor levels of Flupirtine in the brain are almost as high as in the plasma (Obermeier et al. 1985). Hence, the protective doses determined in the present study in vitro are only slightly higher than those needed in patients. However, considering the facts that (1) for the induction of neurotoxicity in vitro approximately 10- to 300-fold higher Glu or NMDA concentrations (1 mM) are required (Erdö and Schäfer 1991) than those measured in the cerebrospinal fluid (10–20 µM; Baethmann et al. 1993), and (2) the sensitivity of the NMDA receptor to its agonists is strongly dependent on divalent cations (Asher and Nowak 1988), a reliable comparison of the effective concentrations obtained from in vitro versus in vivo studies is not possible. Therefore we have strong reasons to assume that Flupirtine is a promising drug for the treatment of neurodegenerative disorders in general and for AIDS-associated encephalopathia in particular.

Acknowledgments. This work was supported by grants from the Deutsche Forschungsgemeinschaft (Mu 348/11-1) and from BMBF (FKZ: 01 KI 9462).

References

Asher P, Nowak L (1988) The role of divalent cations in the N-methyl-D-aspartate response of mouse central neurones in culture. J Physiol 399: 247–266

Baethmann A, Maier-Hauff K, Staub F, Schneider GH, Rothenfusser W, Kempski O (1993) Glutamate as a mediator of secondary brain damage. In: Kempski O (ed) Glutamate – transmitter and toxin. Zuckschwerdt, München, pp 65–75

Billah MM, Bryant RW, Siegel MI (1985) Lipoxygenase products of arachidonic acid modulate biosynthesis of platelet activating factor (1-O-alkyl-2-acetyl-sn-glycero-3-phoshocholine) by human neutrophils via phospholipase A₂. J Biol Chem 260: 6899–6906

Bormann J (1989) Memantine is a potent blocker of N-methyl-D-aspartate (NMDA) receptor channels. Eur J Pharmacol 166: 591–592

Burton KA (1956) Study of the condition and mechanism of diphenylamine reaction for the colorimetric estimation of deoxyribonucleic acid. Biochem J 62: 315–323

Dreyer EB, Kaiser PK, Offermann, JT, Lipton SA (1990) HIV-1 coat protein neurotoxicity prevented by calcium channel antagonists. Science 248: 364–367

Erdö SL, Schäfer M (1991) Memantine is highly potent in protecting cortical cultures against excitotoxic cell death evoked by glutmate and N-methyl-D-aspartate. Eur J Pharamcol 198: 215–217

Friedel HA, Fitton A (1993) Flupirtine. A review of its pharmacological properties, and therapeutic efficacy in pain patients. Drug 45: 548–569

Giulian D, Vaca K, Noonan CA (1990) Secretion of neurotoxins by mononuclear phagocytes infected with HIV-1. Science 250: 1983–1987

Gougeon ML, Oliver R, Garcia S, Guétard D, Dragic T, Dauget C, Montagnier L (1991) Mise en évidence d'un processus d'engagement vers la mort cellulaire par apoptose dans les lymphocytes de patients infectés par le VIH. CR Acad Sci Paris 312: 529–537

Hahn JS, Aizenman E, Lipton SA (1988) Central mammalian neurons normally resistant to glutamate toxicity are made sensitive by elevated extracellular Ca^{2+}: toxicity is blocked by the N-methyl-D-aspartate agonist MK-801. Proc Natl Acad Sci USA 85: 6556–6560

Hidaka H, Inagaki M, Kawamoto S, Saski Y (1984) Isoquinolinesulfonamides, novel and potent inhibitors of cyclic nucleotide-dependent protein kinase and protein kinase C. Biochemistry 23: 5036–5041

Huettner JE, Bean BP (1988) Block of N-methyl-D-aspartate-activated current by the anticonvulsant MK-801: selective binding to open channels. Proc Natl Acad Sci USA 85: 1307–1311

Johnson JW, Asher P (1987) Glycine potentiates the NMDA response in cultured mouse brain neurons. Nature 325: 529–531

Kljajic Z, Schröder HC, Rottman M, Cuperlovic M, Movesian M, Uhlenbruck G, Gasic M, Zahn RK, Müller WEG (1987) A D-mannose-specific lectin from *Gerardia savaglia* that inhibits nucleocytoplasmic transport of mRNA. Eur J Biochem 169: 97–104

Kornhuber JJ, Bormann W, Retz M, Hübers S, Riederer P (1989) Memantine displaces [^{3}H]MK-801 at therapeutic concentrations in postmortem human frontal cortex. Eur J Pharmacol 166: 589–594

Laurent-Crawford AG, Krust B, Riviere Y, Desgranges C, Müller S, Kieny MP, Dauguet C, Hovanessian AG (1993) Membrane expression of HIV envelope glycoproteins triggers apoptosis in CD4 cells. AIDS Res Hum Retrovir 9: 761–773

Lerea LS, McNamara JO (1993) Ionotropic glutamate receptor subtypes activate c-fos transcription by distinct calcium-requiring intracellular signaling pathways. Neuron 10: 31–41

Lipton SA, Sucher NJ, Kaiser PK, Dreyer EB (1991) Synergistic effects of HIV coat protein and NMDA receptor-mediated neurotoxicity. Neuron 7: 111–118

Lynch MA, Errington ML, Bliss TVP (1989) Nordihydroguaiaretic acid blocks the synaptic component of long-term potentiation and the associated increase of glutamate and arachidonate: an in vivo study in the dentate gyrus of the rat. Neuroscience 30: 693–701

Maier-Lenz H, Thieme G (1983) Kumulationskinetik von Flupirtin (D 9998) bei oraler Applikation. Forschungsber Asta-Medica, Frankfurt

McMahon FG, Arndt WF, Newton JJ, Montgomery PA, Perhach JL (1987) Clinical experience with Flupirtine in the U.S. Postgrad Med J 63: 81–85

Miller B, Sarantis M, Traynelis SF, Attwell D (1992) Potentiation of NMDA receptor currents by arachidonic acid. Nature 355: 722–725

Müller WEG, Renneisen K, Kreuter MH, Schröder HC, Winkler I (1988) The D-mannose-specific lectin from *Gerardia savaglia* blocks binding of human immunodeficiency virus type 1 to H9 cells and human lymphocytes in vitro. J Acquir Immune Def Syndr 1: 453–458

Müller WEG, Schröder HC, Ushijima H, Dapper J, Bormann J (1992) Gp120 of HIV-1 induces apoptosis in rat cortical cell cultures: prevention by memantine. Eur J Pharmacol (Mol Pharmacol Sec) 226: 209–214

Obermeier K, Niebich G, Thiemer K (1985) Untersuchungen zur Pharmakokinetik und Bio-transformation des Anagetikums Flupirtin bei Ratte und Hund. Arzneim Forsch Drug Res 35: 60–67

Olney JW (1993) Excitatory transmitter neurotoxicity: an overview. In: Kempski O (ed) Glutamate – transmitter and toxin. Zuckschwerdt, München, pp 1–11

Osborne NN, Pergande, G, Block F, Schwarz M (1994) Immunohistochemical evidence for Flupirtine acting as an antagonist on the N-methyl-D-aspartate (NMDA) and homocysteic acid-induced release of GABA in the rabbit retina. Brain Res 667: 291–294

Perovic S, Schleger C, Pergande G, Iskric S, Ushijima H, Rytik P, Müller WEG (1994) The triaminopyridine Flupirtine prevents cell death in rat cortical cells induced by N-methyl-D-aspartate and gp120 of HIV-1. Eur J Pharmacol (Mol Pharmacol Sec) 288: 27–33

Sanfeliu C, Hunt A, Patel AJ (1990) Exposure to N-methyl-D-aspartate increases release of arachidonic acid in primary cultures of rat hippocampal neurons and not in astrocytes. Brain Res 529: 241–248

Savio T, Levi G (1993) Neurotoxicity of HIV coat protein gp120, NMDA receptors and protein kinase C: a study with cerebellar granule cell cultures. J Neurosci Res 34: 265–272

Schwarz M, Block F, Pergande G (1994) NMDA-mediated muscle relaxant action of Flupirtine in rats. Neuro Rep 5: 1981–1984

Sladeczek G, Pin JP, Recasens M, Bockaert J, Weiss S (1985) Glutamate stimulates inositol phosphate formation in striatal neurons. Nature 317: 717–719

Sung K, Medelow D, Georgescu HI, Evans CH (1988) Characterization of chondrocyte activation in response to cytokines synthesized by a synovial cell line. Biochim Biophys Acta 971: 148–156

Szelenyi I, Nickel B, Borbe HO, Brune K (1989) Mode of antinociceptive action of Flupirtine in the rat. Brit J Pharmacol 97: 835–842

Tapia-Arancibia L, Rage F, Recasens M, Pin JP (1992) NMDA receptor activation stimulates phospholipase A$_2$ and somatostatin release from rat cortical neurons in primary cultures. Eur J Pharmacol (Mol Pharmacol Sect) 225: 253–262

Ushijima H, Kunisada T, Schröder HC, Klöcking P, Müller WEG (1993) HIV 1-gp120 and NMDA differentially induce protein kinase C translocation in rat primary neuronal cultures. J Acquir Immune Def Syndr 6: 339–343

Watkins JC, Evans RH (1981) Excitatory amino acid transmitters. Annu Rev Pharmacol Toxicol 21: 165–204

Weiler BE, Schröder HC, Stefanovich V, Stewart D, Forrest JMS, Allen LB, Bowden BJ, Kreuter MH, Voth R, Müller WEG (1990) Sulphoevernan, a polyanionic polysaccharide, and the Narcissus lectin potently inhibit human immunodeficiency virus infection by binding to viral envelope protein. J Gen Virol 71: 1957–1963

Williams GT (1991) Programmed cell death: apoptosis and oncogenesis. Cell 65: 1097–1098

Wong-Staal F, Gallo RC (1985) Human T-lymphotropic retroviruses. Nature 317: 395–403

Wu FY, Chang NT, Chen WJ, Juan CC (1993) Vitamin K3-induced cell cycle arrest and apoptotic cell death are accompanied by altered expression of c-*fos* and c-*myc* in nasopharyngeal carcinoma cells. Oncogene 8: 2237–2244

Apoptosis of Mature T Lymphocytes: Putative Role in the Regulation of Cellular Immune Responses and in the Pathogenesis of HIV Infection

D. Kabelitz[1], T. Pohl[1], H.H. Oberg[1], K. Pechhold[1], T. Dobmeyer[2], and R. Rossol[2]

Abstract

In this chapter, some aspects of programmed cell death, or apoptosis, of T lymphocytes are discussed. It has been recognized that transformed T cells and immature T lymphocytes can be triggered to undergo apoptosis. As in other cell systems, apoptosis is characterized by cell shrinkage, nuclear condensation, and DNA fragmentation that displays the characteristic "ladder" pattern of approximately 180–200 bp fragments. More recently, however, it has become clear that apoptosis is not restricted to immature thymocytes or transformed T lymphocytes, but can also occur in mature peripheral T cells. This raises the question of whether apoptosis plays a role as a mechanism in regulating cellular immune responses, which will be discussed in the following sections. We will also address the issue of the potential role of T cell apoptosis in pathophysiology. Here, we will concentrate on the infection with human immunodeficiency virus (HIV), where apoptosis is thought to contribute to the continuous decline in CD4[+] T cells.

1 Apoptosis of Immature T Lymphocytes

T lymphocytes are the effector cells of cellular immune responses. On the basis of cell surface markers and/or function, mature T cells can be divided into various subpopulations, of which the CD4[+] and CD8[+] subsets are perhaps the most important. CD4[+] T cells recognize antigen in the context of major histocompatibility complex (MHC) class II molecules and are usually "helper" cells, whereas CD8[+] cells recognize antigen in the context of MHC class I molecules, and are usually "cytotoxic" or "suppressor" cells. In order to recognize antigen, a T cell must express on its surface a T cell receptor (TCR) molecule which is noncovalently associated with the CD3 molecular complex (Ashwell and Klausner 1990).

T lymphocytes acquire TCR molecules during intrathymic T cell differentiation. Immature precursor cells enter the thymus, where the rearrangement of TCR genes and positive and negative selection of developing thymocytes take place (Robey and Fowlkes 1994).

[1] Paul-Ehrlich-Institute, Paul-Ehrlich-Str.51-59 63225 Langen, Germany
[2] Dept. of Hematology, University of Frankfurt, Theodor-Stern-Kai 7, 60590 Frankfurt, Germany

The majority of T cells generated in the thymus express useless or potentially self-reactive TCR molecules. If such cells were allowed to leave the thymus and enter the periphery, autoimmune reactivity would ensue. In fact, the vast majority of thymocytes dies within the thymus; this process is known as "negative selection" (von Boehmer 1992). The intrathymic cell death of large numbers of thymocytes is an active process which is associated with the characteristic hallmarks of apoptosis, and can be mimicked in vitro by antibodies to the CD3/TCR complex (Smith et al. 1989). Apoptosis of immature thymocytes can be triggered not only by monoclonal antibodies (mAb) to CD3/TCR, but also by more physiological ligands such as *Staphylococcus aureus* enterotoxin "superantigens" (Jenkinson et al. 1989). Superantigens stimulate all T cells expressing a particular variable element of the TCR β chain (Vβ) (Marrack and Kappler 1990). Consequently, superantigens eliminate (by apoptosis) in the thymus all developing thymocytes that express the reactive TCR Vβ elements. Moreover, endogenous retroviral superantigens such as the mouse mammary tumor virus (MMTV) also trigger intrathymic apoptosis of thymocytes expressing the reactive TCR Vβ element on the surface (Acha-Orbea et al. 1991).

In addition to immature thymocytes, apoptosis has been triggered in transformed T lymphocytes. Antibodies directed at the CD3/TCR complex activate resting T lymphocytes, i.e., they induce Ca^{2+} influx, cytokine production, and cellular proliferation. The same signals, however, induce growth arrest of activated T cells and transformed T lymphocytes (Breitmeyer et al. 1987; Ucker et al. 1989). Growth arrest was found to be associated with cell shrinkage and fragmentation of DNA into oligonucleosomal bands, and thus displayed characteristic features of apoptosis (Mercep et al. 1989; Takahishi et al. 1989; Odaka et al. 1990).

More recently, however, it became clear that programmed cell death is not restricted to immature thymocytes or transformed T cell lines, but can also be initiated under certain conditions in mature peripheral T lymphocytes (Newell et al. 1990; Lenardo 1991; Russell et al. 1991, 1992; Janssen et al. 1992; Damle et al. 1993a; Wesselborg et al. 1993a). This raises the question as to whether apoptosis of peripheral T cells contributes to the regulation of cellular immune responses.

2 Apoptosis of Mature T Lymphocytes

Cell death can be triggered in activated mature T cells by monoclonal anti-CD3/TCR as well as by mitogenic lectins such as PHA. This phenomenon is observed in both mouse and human T cells of various phenotypes (Russell et al. 1991, 1992; Janssen et al. 1992; Wesselborg et al. 1993a). Apoptosis can also be initiated by signaling via the CD2 molecule (Wesselborg et al. 1993b). It is quite clear that resting T cells are largely resistant to apoptosis induction; only after activation will mature T cells acquire their sensitivity to apoptosis-inducing signals (Russell et al.1991, Klas et al. 1993; Radvanyi et al. 1993; Wesselborg and Kabelitz 1993; Wesselborg et al. 1993a). This seems to ensure that resting T cells, upon encounter of their relevant antigen, can be activated, and will then proliferate as required for the generation of an efficient cellular immune response. It is only after activation (and perhaps after passage through several rounds of the cell cycle) that the T cell might undergo apoptosis upon recognition of antigen; this might then help to terminate an ongoing immune response (Kabelitz et al. 1993). The molecular basis for the lag period required before a T cell becomes susceptible to apoptosis is not well understood. It might be influenced by the regulated expression of gene products such as *fas* and *bcl-2*. *fas*, a member of the TNF-receptor gene family, is an important target antigen for apoptosis induction (Yonehara et al. 1989). Although resting T cell express *fas*, T cells express increased levels of the *fas* antigen after activation (Miyawaki et al. 1992). In line with *fas* antigen expression, resting T cells are quite resistant to apoptosis triggered by anti-*fas* antibody, whereas activated T cells rapidly undergo apoptosis upon *fas*-signaling (Owen-Schaub et al. 1992; Klas et al. 1993).

The *fas* antigen is also expressed on murine cells. There are mouse strains available that have a genetic defect in either *fas* antigen (*lpr* strain) or the *fas* ligand (*gld* strain) expression. Both mouse strains suffer from a generalized lymphoproliferative disease which is interpreted to result from defective apoptosis signaling and, therefore, decreased elimination of lymphocytes (Watanabe-Fukunaga et al. 1992; Takahashi et al. 1994). Interestingly, both mouse strains have a defect in the apoptosis pathway mediated by anti-CD3/TCR antibodies. This may be taken as evidence that the CD3/TCR-dependent death pathway requires a functional *fas/fas* ligand expression (Bossu et al. 1993; Russell and Wang 1993; Russell et al. 1993).

Another gene product which appears to be involved in the regulation of apoptosis is the *bcl-2* proto-oncogene. *bcl-2* seems to protect the cell from apoptosis, possibly by an anti-oxidative mechanism (Hockenbery et al. 1993). *bcl-2* appears to conteract the *fas*-induced apoptosis (Itoh et al. 1993). Moreover, *bcl-2*-deficient mice generated by homologous recombination suffer from fulminant lymphoid apoptosis (Veis et al. 1993).

It is clear, however, that apoptosis is a complex biological process that involves the activation of many genes. Apoptosis contributes in a controlled manner to the morphogenesis of *Caenorhabditis elegans*, and some of the genes defined in *C. elegans* have been found to have homologues in mouse and/or human cells (Schwartz and Osborne 1993). Genes involved in the regulation of the cell cycle such as the cyclin-dependent kinases, are also important in the regulation of apoptosis (Schwartz and Osborne 1993; Shi et al. 1994). In T lymphocytes, sensitivity to apoptosis appears to be highest during the S phase, whereas cells arrested in the G1 phase are largely resistant (Boehme and Lenardo 1993).

3 Apoptosis of T Lymphocytes Induced by Superantigens and Conventional Antigen

As discussed above, superantigens are potent inducers of apoptosis in thymocytes. When a superantigen such as SEB is injected into mice, T cells expressing

Vβ 8 are activated and expanded in vivo. The number of Vβ 8 T cells drops below the initial number, however, after several days the specific loss of peripheral Vβ 8 T cells is due to apoptosis (Kawabe and Ochi 1991; MacDonald et al. 1991; Gonzalo et al. 1992; Lussow et al. 1993). Superantigens are also useful tools in investigating apoptosis of mature T lymphocytes in vitro. Activated mature T cells undergo apoptosis upon contact with superantigen presented by MHC class II molecules, provided that they express the required TCR Vβ element (Damle et al. 1993b; Wesselborg et al. 1993a). We have used human SEE-reactive T cell clones to investigate the role of antigen-presenting cells (APC) in the induction (or prevention) of enterotoxin-induced T cell apoptosis (Kabelitz and Wesselborg 1992). SEE killed a major proportion (40–60%) of reactive cells (even within a given T-cell clone), independently of whether APC were present or not. Importantly, however, a proliferative response of clone cells to SEE stimulation was observed only if APC were present. Thus, it appeared that APC could not prevent apoptosis of T cells, but were required to initiate a proliferative response in the surviving fraction of the clone T cells (Kabelitz and Wesselborg 1992). The exact role of APC in the regulation of superantigen-induced T-cell death is not clear and needs further investigation. There is evidence, however, that cell interaction molecules such as CD11a/CD18 play a role (Damle et al. 1993b).

Although superantigens have been convincingly shown to induce apoptosis of activated mature T lymphocytes, it is less clear whether programmed cell death can also be triggered by conventional antigens. Recently, however, evidence to support this contention has been obtained (Kabelitz et al. 1995). MHC class I-expressing cytotoxic T cells (CTL) can present antigenic peptides to each other; this triggers self-destruction of murine and human CTL (Walden and Eisen 1990; Suhrbier et al. 1993). Antigen-derived peptides trigger cell death not only in vitro but also in vivo, as has been shown in T cell receptor transgenic mice where injection of the relevant peptide induces deletion of transgenic T cells which carry the TCR with specificity for that particular peptide (Kyburz et al. 1993; Mamalaki et al. 1993). At least under in vitro conditions, both CD4$^+$ and CD8$^+$ T-cell subsets are susceptible to programmed cell death. It is interesting to note that activated HLA class II-reactive human CD4$^+$ T cells undergo apoptosis when recognizing APC expressing the relevant HLA-DR molecules (Damle et al. 1993a). Moreover, murine CD8$^+$ MHC class I-specific T cells undergo apoptosis upon recognition of the appropriate MHC class I alloantigen (Ucker et al. 1992).

To investigate antigen-induced death of human T lymphocytes, we have developed a novel flow cytometry method termed standard cell dilution assay (SCDA). This method allows a quantitative determination of the absolute number of viable cells of any given lymphocyte subset within a heterogeneous cell population (Pechhold et al. 1994). When applying SCDA to the analysis of activation-induced T cell death in vitro, the reduction of the number of viable responder T cells after stimulation with the specific or with unrelated antigens is measured. When alloantigen-reactive human T cells that have been generated by repeated restimulation are activated by the specific stimulator cells, 20–30% of the responding T cells die during overnight culture; under these conditions, both

CD4[+] and CD8[+] T cells are susceptible to activation-induced cell death (Kabelitz et al. 1994). Even though significant levels of T cell death occur following restimulation with alloantigen, essentially no fragmentation of DNA into low molecular weight repeats can be detected (Pohl et al., unpubl. observ). This illustrates that activation-dependent cell death can proceed in the absence of significant DNA fragmentation (Schulze-Osthoff et al. 1994). It is possible that the "intensity" of TCR signaling determines whether or not signal-dependent T cell death is associated with DNA fragmentation. While mitogenic lectins (such as PHA) and superantigens are "strong" activators, stimulations through recognition of allogeneic MHC molecules is presumably less intense. However, alternative explanations can be envisaged, and further investigations at the biochemical level of signal transduction are required to clarify this issue. Induction of T cell death as revealed by SCDA is not restricted to alloantigen-reactive T cells but is similarly observed upon restimulation of tetanus toxoid-specific CD4[+] T lymphocytes with the specific but not with unrelated antigen (Kabelitz et al. 1994). Thus, it appears to be a general phenomenon that activated mature peripheral T lymphocytes die upon contact with nonspecific stimuli (PHA, anti-TCR/CD3 antibodies, etc.) or specific (super)antigen. This raises the question of whether activation-induced cell death is a physiological process that contributes to the regulation of cellular immune responses under normal conditions (Kabelitz et al. 1993). On the basis of this assumption, it can be postulated that altered levels of T-cell death in pathophysiological situations are likely to result in immunodeficiency (enhanced T-cell apoptosis) or autoimmunity (descreased T-cell apoptosis).

4   Apoptosis of Mature T Lymphocytes In Vivo and Ex Vivo

What is the evidence that apoptosis of mature T cells plays any role in vivo? We would like to discuss the following two examples. Experimental allergic encephalomyelitis (EAE) is an autoimmune disease which is mediated by CD4 T cells with specificity for myelin basic protein (MBP). The clinical symptoms are due to the demyelinization resulting from the attack by MBP-specific CD4 T cells. Pender et al. (1992) obtained evidence by ultrastructural analysis that a proportion of CD4[+] T cells infiltrating the parenchyma of the spinal cord underwent apoptosis in situ; similar data were obtained by Schmied et al. (1993).

This observation prompted Critchfield and coworkers (1994) to explore the possibility of inducing apoptosis of pathogenic T cells by high concentrations of antigen. In fact, the injection of high concentrations of MBP deleted MBP-reactive CD4[+] T cells in vivo and reduced the clinical and pathological signs of autoimmune disease. Further experiments with other models of autoimmune disease will show whether it is a generally applicable approach to delete (auto)-antigen-reactive T cells in vivo by high concentrations of antigen.

A second example of T-cell death occurring in vivo (or ex vivo) is viral infection. Activation of T lymphocytes during the process of viral infection seems to predispose T cells to subsequent apoptosis initiated by signaling through the

TCR/CD3 molecular complex (Moss et al. 1985; Uehara et al. 1992; Razvi and Welsh 1993; Tamaru et al. 1993) Acute infection with Epstein-Barr virus (infectious mononuclosis) is characterized by a lymphocytosis and the appearance of lymphoblasts in the peripheral blood; activated lymphoblasts are EBV-specific CD8[+] T cells. Upon transfer into in vitro culture, a large proportion of lymphoblasts undergoes spontaneous apoptosis (Uehara et al. 1992). These data suggest that programmed cell death after massive immune activation following acute virus infection is a physiological means to terminate ongoing immune responses.

5 Role of T-Cell Apoptosis in HIV Pathogenesis

During progression from the asymptomatic stage of human immunodeficiency virus (HIV) infection to AIDS, a continuous decline in CD4[+] T lymphocytes takes place. A number of mechanisms, which are not mutually exclusive, has been discussed as contributing to the decline in CD4[+] T-lymphocytes: (1) HIV-induced cytolysis, directly or through syncytia formation (Tersmette and Schuitemaker 1993). (2) HIV-induced dysregulation of cytokines (Rossol et al. 1989; Rossol-Voth et al. 1990; Shearer and Clerici 1991; Meyaard et al. 1993), (3) HIV-induced autoimmune diseases (Leelayuwat et al. 1993; Ditzel et al. 1994), (4) cytotoxic T-lymphocyte response (Zinkernagel and Hengartner 1994; Clerici et al. 1993; Rossol-Voth et al. 1988); recent reports have suggested that programmed cell death of CD4[+] T cells may ultimately lead to the depletion of CD4[+] T lymphocytes during HIV infection (Ameisen 1992; Gougeon and Montagnier 1993; Ameisen 1994; Finkel and Banda 1994).

There is some controversy as to whether T cells from HIV-infected individuals undergo enhanced "spontaneous" apoptosis in vitro when compared to HIV-negative donors. While some reports have shown that this is indeed the case (Meyaard et al. 1992; Gougeon et al. 1993), others have observed only very moderate levels of spontaneous T-cell death when peripheral blood lymphocytes from HIV[+] donors were cultured without any stimulus (Groux et al. 1992). In our own studies, we did not detect a significant "spontaneous" reduction in CD4[+] T cells from HIV[+] individuals within 48 h of in vitro culture (Wesch et al., unpubl.). It is quite clear, however, that T cells from HIV[+] individuals show enhanced apoptosis compared to healthy controls when they are activated by mitogens or superantigens (Groux et al. 1992; Meyaard et al. 1992, 1994; Finkel and Banda 1994). Interestingly, enhanced in vitro apoptosis is not restricted to the CD4[+] T cell subset but also affects CD8[+] T cells (Meyaard et al. 1992). Recent evidence suggests that CD8[+] T cell apoptosis is not specific to the pathogenesis of AIDS, but presumably reflects the massive immune activation discussed above in the context of other viral infections. In contrast, CD4[+] T cell apoptosis seems to be specific to the pathogenesis of AIDS, because it is found in pathogenic but not in nonpathogenic lentivirus infections (Estaquier et al. 1994).

The binding of CD4 coreceptors to the external glycoprotein gp120 of the virus primes T cells for programmed cell death (Banda et al. 1992). In combination with additional T-cell receptor stimulation, T lymphocytes then undergo apoptosis. Moreover, soluble gp120 and membrane-bound gp120 expressed on HIV-infected cells, which come into direct contact with noninfected cells, may then be able to prime noninfected T cells for apoptosis. gp120 was the first, but is perhaps not the only candidate for the induction of apoptosis in HIV infection. Interestingly, stimulatory consignals, i.e., anti CD28 monclonal antibodies or the cytokines interleukin-1 and interleukin-2 prevent gp120-mediated apoptosis in T cells from healthy controls as well as from HIV-infected individuals (Amendola et al. 1994).

Additional mechanisms may contribute to apoptosis in HIV infection (Kornbluth 1994). Because of the existence of superantigens in animal retroviruses, it has been suggested that HIV might also encode a superantigen. However, data providing evidence for this hypothesis are not yet available.

Furthermore, altered antigen-presenting cell function may account for programmed cell death in HIV infection. Monocytes, macrophages, Langerhans cells, and probably dendritic cells are susceptible to infection with HIV (Crowe and Kornbluth 1994; Schuitemaker 1994). Upon HIV infection, these cells show an altered expression of accessory cell surface molecules and an altered release of cytokines. In addition, decreased accessory cell function could be demonstrated in monocytic cells lines after in vitro infection with HIV. In fact, the ability of mature T cells to tolerate self and to react against nonself does not depend solely on the nature of the T-cell receptor, but also on cosignals delivered by antigen-presenting cells. T-cell receptor stimulation plus a costimulatory second signal induces T cell proliferation and differentiation into effector and memory cells, whereas T cell receptor stimulation alone may induce anergy and - alternatively or consequently – apoptosis (Linsley and Ledbetter 1993). Moreover, one may speculate that a molecular mimicry between HIV gp160 and MHC class II molecules may cause an inappropriate tolerance to self-MHC II restricted signals. Any isolated activation signal may prime cells for apoptosis, whereas an intact microenvironment and a certain "mixture" of multiple signals may be necessary to induce cell proliferation. The demonstration that the addition of anti CD28 antibody or interleukin-2 could block anti-CD3-induced programmed cell death, at least under certain conditions, provides support for this hypothesis (Meyaard et al. 1993; Leelayuwat et al. 1993).

In addition, many agents which induce apoptosis are either oxidants or stimulators of cellular oxidative metabolism. Conversely, many inhibitors of apoptosis have antioxidant activites. The pivotal role of reactive oxygen intermediates for apoptosis has recently been elucidated by the demonstration that the proto-oncogene *bcl-2* prevents apoptosis in an antioxidant manner (Hockenbery et al. 1993). As the terminal electron acceptor for oxidative phosphorylation, molecular oxygen plays a crucial role in all aerobic forms of life. Yet, it is also toxic under certain conditions, and ultimately, it will oxidize all biological matter. The use of molecular oxygen is associated with the formation of a variety of reactive oxygen intermediates, oxygen species which will easily react with cellular macromolecules resulting in extensive damage of cellular structures. Phagocytic cells take advantage of this fact and most of their antimicrobial activities can be attributed to the formation of reactive oxygen species by a membrane bound enzyme. The dichotomy of the paradoxical need for an ultimately toxic substance is called oxygen paradox. Therefore, all aerobic cells are in a persistent state of oxidative stress. On the other hand, cells elaborate adaptive responses in order to detoxify reactive oxygen intermediates. Indeed, cells synthesize chemicals and enzymes to protect themselves from the influence of reactive oxygen species.

The process of apoptosis may be linked to oxidative stress in HIV infection. An excessive production of reactive oxygen species and a broad deficiency of antioxidant mechanisms provide support for this hypothesis. As reactive oxygen species are a physiological product of phagocytic cells, any antigenic or regulatory stimulus to phagocytes, as it occurs during opportunistic infections, will increase the generation of reactive oxygen products. Furthermore, macrophages and monocytes produce TNF-alpha, which is found in increased concentrations in patients infected with HIV. In addition, TNF enhances the respiratory burst of monocytes and neutrophils. Moreover, patients infected with HIV demonstrate not only an enhanced generation of reactive oxygen intermediates, but also show a reduction of antioxidant defense mechanisms. The oxygen radical scavengers superoxide dismutase, catalase, and glutathione are found to be reduced in HIV-infected subjects. Gut infections and malabsorption lead to a reduced uptake of fat-soluble antioxidants, including vitamins, flavonoids, and quinones. Furthermore, increased serum concentrations of TNF cause malabsorption and a reduced nutritional supply with antioxidative compounds. All these mechanisms results in a fatal imbalance between reactive oxygen intermediates (ROI) production on the one hand and antioxidative capacities on the other (Buttke and Sandstrom 1994; Greenspan and Aruoma 1994).

In what way might oxygen radicals mediate apoptosis? Direct reactive oxygen species-mediated DNA damage has been demonstrated to induce the activation of poly ADP-ribose transferase and of p53 – both strongly associated with apoptosis. The polymerization of ADP-ribose to proteins induces the consumption of NAD/NADH. The loss of NAD, which is absolutely necessary for ATP synthesis, leads to a reduction of cellular ATP. The diminished concentration of NADH as a reduction equivalent results in a deficiency of glutathione in its reduced form. In addition, the oxidation of membrane lipids leads to the formation of metabolites, which are potent inducers of apoptosis and which, in addition, enhance ROI generation (Bäuerle and Baltimore 1988; Shaposhnikova et al. 1994). Furthermore, the transcription factor NFkB (and probably others) have been shown to be under redox control. Reactive oxygen intermediates have been demonstrated to stimulate the release of NFkB from its inhibitory subunit; NFkB might then be able to induce the transcription of several genes, which will then cause apoptosis (Bäuerle and Baltimore 1988). T cells could be triggered to undergo apoptosis by passage through reactive oxygen species-enriched tissues such as lymph nodes, where T cells come into direct contact with reactive oxygen species producing macrophages (Buttke and Sandstrom 1994; Greenspan and Aruoma 1994).

Stimulation of the TNF receptor results in a rapid rise in apoptosis. The exact process of TNF-mediated apoptosis is still unknown, but recent data suggest that redox mechanisms play an important role. Interestingly, TNF-induced apoptosis may be inhibited either by thioredoxin, an intracellular thiol reductant and free radical scavenger, or N-acetylcystein, a thiol antioxidant and glutathione precursor (Mayer and Noble 1994). Moreover, cellular resistance to TNF is correlated with decreased or increased concentrations of superoxide dismutase. To investigate TNF-α and oxygen radical-induced apoptosis of human lymphocytes, we coincubated mononuclear cells isolated from healthy individuals, patients infected with HIV, and human unbilical cord blood with various doses of TNF-α and with reactive oxygen intermediates generated by monocytes during phagocytosis-induced respiratory burst. It is important to mention that under certain conditions, CD4$^+$ lymphocytes are more susceptible to TNF-α and oxidative stress-induced programmed cell death than other lymphocyte subsets including T and B cells, indicating some kind of priming for apoptosis (Rossol et al., unpubl. observ.).

Clarifying the role and the mechanisms of apoptosis could be important for therapeutic intervention in HIV infection. However, many questions still remain to be answered. Thus, it is unclear, why CD4$^+$ but not CD8$^+$ T cells decline preferentially during HIV infection. Perhaps CD8$^+$ cells also die in vivo but are efficiently replaced. On the other hand, it is possible that CD4$^+$ T-cell depletion is a consequence of an impairment of T cell renewal rather than of a depletion. Furthermore, one may speculate that the Th1-Th2 shift, which is discussed as, occurring during HIV infection (Clerici and Shearer 1993), is due to programmed cell death of a certain T-helper cell subtype. The preferential induction of Th1 apoptosis may be due to altered antigen presenting cells stimulating different Th subsets. Finally, one has to consider that apoptosis in HIV infection may serve as a physiological rather than a pathophysiological mechanism to eliminate certain "pathological" cells. Preventing apoptosis would then be counterproductive in the treatment of HIV infection.

6 Concluding Remarks

Apoptosis as a mechanism of signal-dependent active ("suicidal") form of cell death usually associated with distinct DNA fragmentation has been known for a long time to play an important role in organogenesis and differentiation of multicellular organisms (Kerr et al. 1972). However, it was only a few years ago that immunologists recognized the role of apoptosis in the immune system. It appears that programmed cell death controls the homoeostasis of the immune system, from the intrathymic deletion of unwanted immature thymocytes to the regulation of cellular immune reactions in the periphery. A precise understanding of the molecular mechanism of T-lymphocyte apoptosis might provide a basis for developing rational strategies of inducing or preventing apoptosis under pathophysiological conditions.

Acknowledgments. Work from our laboratory was generously supported by the Alfried Krupp Award for young professors (to D.K.). We would like to thank Constanze Taylor for expert secretarial assistance.

References

Acha-Orbea H, Shakhov AN, Scarpellino L, Kolb E, Müller V, Vessaz-Shaw A, Fuchs R, Bölchlinger K, Rollini P, Billotte J, Sarafidou M, MacDonald HR (1991) Clonal delection of Vβ 14-bearing T cells in mice transgenic for mammary tumour virus. Nature 350: 207–211

Ameisen JC (1992) Programmed cell death and AIDS: from hypothesis to experiment. Immunol Today 13: 388–391

Ameisen JC (1994) Programmed cell death (apoptosis) and cell survival regulation: relevance to AIDS and cancer. AIDS 8: 1197–1213

Amendola A, Lombardi G, Oliverio S, Colizzi V, Piacentini M (1994) HIV-1 gp120-dependent induction of apoptosis in antigen-specific human T cell clones is characterized by tissue transglutaminase expression and prevented by cyclosporin A. FEBS Lett 339: 258–264

Ashwell JD, Klausner RD (1990) Genetic and mutational analysis of the T-cell antigen receptor. Annu Rev Immunol 8: 139–167

Banda NK, Bernier J, Kurahara DK, Kurrle R, Haigwood N, Sekaly RP, Finkel TH (1992) Crosslinking CD4 by human immunodeficiency virus gp120 primes T cells for activation-induced apoptosis. J Exp Med 176: 1099–1106

Bäuerle PA, Baltimore D (1988) 1 kappa B: a specific inhibitor of the NF kappa B transcription factor. Science 242: 540–546

Boehme SA, Lenardo MJ (1993) Propriocidal apoptosis of mature T lymphocytes occurs at S phase of the cell cycle. Eur J Immunol 23: 1552–1560

Bossu P, Singer GG, Andres P, Ettinger R, Marshak-Rothstein A, Abbas AK (1993) Mature CD4$^+$ T lymphocytes from MRL/lpr mice are resistant to receptor-mediated tolerance and apoptosis. J Immunol 151: 7233–7239

Breitmeyer JB, Oppenheim SO, Daley JF, Levine HB, Schlossman SF (1987) Growth inhibition of human T cells by antibodies recognizing the T cell antigen receptor complex. J Immunol 138: 726–731

Buttke TM, Sandstrom PA (1994) Oxidative stress as a mediator of apoptosis. Immunol Today 15: 7–10

Clerici M, Shearer GM (1993) A Th1 to Th2 switch is a critical step in the etiology of HIV infection. Immunol Today 14: 107–111

Clerici M, Shearer GM, Housell EF, Jameson B, Habeshaw J, Dalgleish AG (1993) Alloactivated cytotoxic T cells recognize the carboxy-terminal domain of human immunodeficiency virus-1 gy120 envelope glycoprotein Eur J Immunol 23: 2022–2025

Critchfield JM, Racke MK, Zùñiga-Pflücker JC, Cannella B, Raine CS, Goverman J, Lenardo MJ (1994) T cell deletion in high antigen does therapy of autoimmune encephalomyelitis. Science 263: 1139–1242

Crowe SM, Kornbluth RS (1994) Overview of HIV interactions with macrophages and dendritic cells: the other infection in AIDS. J Leukoc Biol 56: 215–217

Damle NK, Klussman K, Leytze G, Aruffo A, Linsley PS, Ledbetter JA (1993a) Costimulation with integrin ligands intercellular adhesion molecule-1 or vascular cell adhesion molecule-1 augments activation-induced death of antigen-specific CD4$^+$ T lymphocytes. J Immunol 151: 2368–2379

Damle NK, Leytze G, Klussman K, Ledbetter JA (1993b) Activation with superantigens induces programmed death in antigen-primed CD4$^+$ class II$^+$ major histocompatibility complex T lymphocytes via a CD11a/CD18-dependent mechanism. Eur J Immunol 23: 1513–1522

Ditzel HJ, Barbas SM, Barbas CF, Burton DR (1994) The nature of the autoimmune antibody repertoire in human immunodeficiency virus type 1 infection. Proc Natl Acad Sci USA 91: 3710–3714

Estaquier J, Idziorek T, de Bels F, Barré-Sinoussi F, Hurtrel B, Aubertin AM, Venet A, Mehtali J, Muchmore E, Michel P, Mouton Y, Girard M, Ameisen JC (1994) Programmed cell death and AIDS: significance of T-cell apoptosis in pathogenic and nonpathogenic primate lentiviral infections. Proc Natl Acad Sci USA 91: 9431–9435

Finkel TH, Banda NK (1994) Indirect mechanisms of HIV pathogenesis: how does HIV kill T cells? Curr Opin Immunol 6: 605–615

Gonzalo JA, de Alborán IM, Alés-Martínez JE, Martínez-AC, Kroemer G (1992) Expansion and clonal deletion of peripheral T cells induced by bacterial superantigen is independent of the interleukin-2 pathway. Eur J Immunol 22: 1007–1011

Gougeon ML, Garcia S, Heeney J, Tschopp R, Lecoeur H, Guetard D, Rame V, Dauguet C, Montagnier L (1993) Programmed cell death in AIDS-related HIV and SIV infections. AIDS Res Hum Retrovir 9: 553–563

Gougeon ML, Montagnier L (1993) Apoptosis in AIDS, Science 260: 1269–1270

Greenspan HC, Aruoma OI (1994) Oxidative stress and apoptosis in HIV infection: a role for plant-derived metabolites with synergistic antioxidant activity. Immunol Today 15: 209–213

Groux H, Torpier G, Monte D, Mouton Y, Capron A, Ameisen JC (1992) Activation-induced death by apoptosis in CD4$^+$ T cells from human immunodeficiency virus-infected asymptomatic individuals. J Exp Med 175: 331–340

Hockenbery DM, Oltvai ZN, Yin X-M, Milliman CL, Korsmeyer SJ (1993) Bcl-2 functions in an antioxidant pathway to prevent apoptosis. Cell 75: 241–251

Itoh N, Tsujimoto Y, Nagata S (1993) Effect of bcl-2 on fas antigen-mediated cell death. J Immunol 151: 621–627

Janssen O, Wesselborg S, Kabelitz D (1992) The immunosuppressive action of OKT3. OKT3 induces programmed cell death (apoptosis) in activated human T cells. Transplantation 53: 233–234

Jenkinson EJ, Kingston R, Smith CA, Williams GT, Owen JJT (1989) Antigen-induced apoptosis in developing T cells: a mechanism for negative selection of the T cell receptor repertoire. Eur J Immunol 19: 2175–2177

Kabelitz D, Wesselborg S (1992) Life and death of a superantigen-reactive human CD4$^+$ T cell clone: staphylococcal enterotoxins induce death by apoptosis but simultaneously trigger a proliferative response in the presence of HLA-DR$^+$ antigen-presenting cells. Int Immunol 4: 1381–1388

Kabelitz D, Pohl T, Pechhold K (1993) Activation-induced cell death (apoptosis) of mature peripheral T lymphocytes. Immunol Today 14: 338–339

Kabelitz D, Oberg HH, Pohl T, Pechhold K (1994) Antigen-induced death of mature T lymphocytes: analysis by flow cytometry. Immunol Rev 142: 157–174

Kabelitz D, Pohl T, Pechhold K (1995) T cell apoptosis triggered via the CD3/T cell receptor complex and alternative activation pathways. Curr Top Microbiol Immunol 200: 1–14

Kawabe Y, Ochi A (1991) Programmed cell death and extrathymic reduction of Vβ8$^+$ CD4$^+$ T cells in mice tolerant to Staphylococcus aureus enterotoxin B. Nature 349: 245–248

Kerr JFR, Wyllie AH, Currie AR (1972) Apoptosis: a basic biological phenomenon with wide-ranging implications in tissue kinetics. Br J Cancer 26: 239–249

Klas C, Debatin KM, Jonker RR, Krammer PH (1993) Activation interferes with the APO-1 pathway in mature human T cells. Int Immunol 5: 625–630

Kornbluth RS (1994) Significance of T cell apoptosis for macrophages in HIV infection. J Leukoc Biol 56: 247–256

Kyburz D, Aichele P, Speiser DE, Hengartner H, Zinkernagel RM, Pircher H (1993) T cell immunity after a viral infection versus T cell tolerance induced by soluble viral peptides. Eur J Immunol 23: 1956–1962

Leelayuwat C, Zhang WJ, Abraham LJ, Townsend DC, Gaudieri S, Dawkins RL (1993) Differences in the central major histocompatibility complex between humans and chimpanzees. Implications for development of autoimmunity and acquired immune deficiency syndrome. Hum Immunol 38: 30–41

Lenardo MJ (1991) Interleukin-2 programs mouse $\alpha\beta$ T lymphocytes for apoptosis. Nature 353: 858–861

Linsley PS, Ledbetter JA (1993) The role of the CD28 receptor during T cell responses to antigen. Annu Rev Immunol 11: 191–212

Lussow AR, Crompton T, Karapetian O, MacDonald HR (1993) Peripheral clonal deletion of superantigen-reactive T cells is enhanced by cortisone. Eur J Immunol 23: 578–581

MacDonald HR, Baschieri S, Lees RK (1991) Clonal expansion precedes anergy and death of Vβ 8⁺ peripheral T cells responding to staphylococcal enterotoxin B in vivo. Eur J Immunol 21: 1963–1966

Mamalaki C, Tanaka Y, Corbella P, Chandler P, Simpson E, Kioussis D (1993) T cell deletion follows chronic antigen-specific T cell activation in vivo. Int Immunol 5: 1285–1292

Marrack P, Kappler J (1990) The staphylococcal enterotoxins and their relatives. Science 248: 705–711

Mayer M, Noble M (1994) N-Acetyl-L-cysteine is a pluripotent protector against cell death and enhancer of trophic factor-mediated cell survival in vitro. Proc Natl Acad Sci USA 91: 7496–7500

Mercep M, Noguchi PD, Ashwell JD (1989) The cell cycle block and lysis of an activated T cell hybridoma are distinct processes with different Ca²⁺ requirements and sensitivity to cyclosporine A. J Immunol 142: 4085–4092

Meyaard L, Otto SA, Jonker RR, Mijnster MJ, Keet RPM, Miedema F (1992) Programmed death of T cells in HIV-1 infection. Science 257: 217–219

Meyaard L, Schuitemaker H, Miedema F (1993) T-cell dysfunction in HIV infection: anergy due to defective antigen-presenting cell function? Immunol Today 14: 161–164

Meyaard L, Otto SA, Keet IPM, Roos MTL, Miedema F (1994) Programmed death of T cells in human immunodeficiency virus infection. No correlation with progression to disease. J Clin Invest 93: 982–988

Miyawaki T, Uehara T, Nibu R, Tsuji T, Yachie A, Yonehara S, Taniguchi N (1992) Differential expression of apoptosis-related fas antigen on lymphocyte subpopulation in human peripheral blood. J Immunol 149: 3753–3758

Moss DJ, Bishop CJ, Burrows SR, Ryan JM (1985) T lymphocytes in infectious mononucleosis. I. T cell death in vitro. Clin Exp Immunol 60: 61–69

Newell MK, Haughn LJ, Maroun CR, Julius MH (1990) Death of mature T cells by separate ligation of CD4 and the T-cell receptor for antigen. Nature 347: 286–288

Odaka C, Kizaki H, Tadakuma T (1990) T cell receptor-mediated DNA fragmentation and cell death in T cell hybridomas. J Immunol 144: 2096–2101

Owen-Schaub LB, Yonehara S, Crump III WL, Grimm E (1992) DNA fragmentation and cell death is selectively triggered in activated human lymphocytes by fas antigen engagement. Cell Immunol 140: 197–205

Pechhold K, Pohl T, Kabelitz D (1994) Rapid quantification of lymphocyte subsets in heterogeneous cell populations by flow cytometry. Cytometry 16: 152–159

Pender MP, McCombe PA, Yoong G, Nguyen KB (1992) Apoptosis of alpha beta T lymphocytes in the nervous system in experimental autoimmune encephalomyelitis: its possible implications for recovery and acquired tolerance. J Autoimmunol 5: 401–410

Radvanyi LG, Mills GB, Miller RG (1993) Religation of the T cell receptor after primary activation of mature T cells inhibits proliferation and induces apoptotic cell death. J Immunol 150: 5704–5715

Razvi ES, Welsh RM (1993) Programmed cell death of T lymphocytes during acute viral infection: a mechanism for virus-induced immune deficiency. J Virol 67: 5754–5765

Robey E, Fowlkes BJ (1994) Selective events in T cell development. Annu Rev Immunol 12: 675–705

Rossol S, Rossol R, Laubenstein HP, Müller WEG, Schröder HC, Meyer zum Büschenfelde KH, Hess G (1989) Interferon production in patients infected with HIV-1. J Infect Dis 159: 815–821

Rossol R, Rossol S, Gräff E, Laubenstein HP, Müller WEG, Meyer zum Büschenfelde KH, Hess G (1988) Natural killer cell activity as a prognostic parameter in the progression to AIDS. J Infect Dis 157: 851–860

Rossol R, Rossol S, Klein K, Hess G, Schütt H, Schröder HC, Meyer zum Büschenfelde KH, Müller WEG (1990) Differential gene expression of IFN-alpha and TNF-alpha in peripheral blood mononuclear cells from patients with AIDS related complex and AIDS. J Immunol 144: 970–975

Russell JH, Wang R (1993) Autoimmune gld mutation uncouples suicide and cytokine/proliferation pathways in activated mature T cells. Eur J Immunol 23: 2379–2382

Russell JH, White CL, Loh DY, Meleedy-Rey P (1991) Receptor-stimulated death pathway is opened by antigen in mature T cells. Proc Natl Acad Sci USA 88: 2151–2155

Russell JH, Rush BJ, Abrams SI, Wang R (1992) Sensitivity of T cells to anti-CD3-stimulated suicide is independent of functional phenotype. Eur J Immunol 22: 1655–1658

Russell Jh, Rush B, Weaver C, Wang R (1993) Mature T cells of autoimmune lpr/lpr mice have a defect in antigen-stimulated suicide. Proc Natl Acad Sci USA 90: 4409 – 4413

Schmied M, Brietschopf H, Gold R, Zischler H, Rothe G, Wekerle H, Lassmann H (1993) Apoptosis of T lymphocytes in experimental autoimmune encephalomyelitis. Evidence for programmed cell death as a mechanism to control inflammation in the brain. Am J Pathol 143: 446–452

Schuitemaker H (1994) Macrophage-tropic HIV-1 variants: initiators of infection and AIDS pathogenesis? J Leukoc Biol 56: 218–224

Schulze-Osthoff K, Walczak H, Dröge W, Krammer PH (1994) Cell nucleus and DNA fragmentation are not required for apoptosis. J Cell Biol 127: 15–20

Schwartz LM, Osborne BA (1993) Programmed cell death, apoptosis and killer genes. Immunol Today 14: 582–590

Shaposhnikova VV, Dobrovinskaya OR, Eidus LK, Korystov YN (1994) Dependence of thymocyte apoptosis on protein kinase C and phospholipase A2. FEBS Lett 348: 317–319

Shearer GM, Clerici M (1991) Early helper defects in HIV infection. AIDS 5: 245–253

Shi L, Nishioka WK, Th'ng J, Bradbury EM, Litchfield DW, Greenberg AH (1994) Premature p34^{cdc2} activation required for apoptosis. Science 263: 1143–1145

Smith CA, Williams GT, Kingston R, Jenkinson EJ, Owen JJT (1989) Antibodies to CD3/T-cell receptor complex induce death by apoptosis in immature T cells in thymic cultures. Nature 337: 181–184

Suhrbier A, Burrows SR, Fernan A, Lavin MF, Baxter GD, Moss DJ (1993) Peptide epitope-induced apoptosis of human cytotoxic T lymphocytes. Implications for peripheral T cell deletion and peptide vaccination. J Immunol 150: 2169–2178

Takahashi T, Tanaka M, Brannan CI, Jenkins NA, Copland NG, Suda T, Nagata S (1994) Generalized lymphoproliferative disease in mice caused by a point mutation in the fas ligand. Cell 76: 969 – 976

Takahishi S, Maecker HT, Levy R (1989) DNA fragmentation and cell death mediated by T cell antigen receptor/CD3 complex on a leukemia T cell line. Eur J Immunol 19: 1911–1919

Tamaru Y, Miyawaki T, Iwai K, Tsuji T, Nibu R, Yachie A, Koizumi S, Taniguchi N (1993) Absence of bcl-2 expression by activated CD45RO$^+$ T lymphocytes in acute infectious mononucleosis supporting their susceptibility to programmed cell death. Blood 82: 521–527

Tersmette M, Schuitemaker H (1993) Virulent HIV strains? AIDS 7: 1123–1125

Ucker DS, Ashwell JD, Nickas G (1989) Activation-driven T cell death. I. Requirements for de novo transcription and translation and association with genome fragmentation. J Immunol 143: 3461–3469

Ucker DS, Meyers J, Obermiller PS (1992) Activation-driven T cell death. II. Quantitative differences alone distinguish stimuli triggering nontransformed T cell proliferation or death. J Immunol 149: 1583–1592

Uehara T, Miyawaki T, Ohta K, Tamaru Y, Yokoi T, Nakamura S, Taniguchi N (1992) Apoptotic cell death of primed CD45RO$^+$ T-lymphocytes in Epstein-Barr virus-induced infectious mononucleosis. Blood 80: 452–458

Veis DJ, Sorenson CM, Shutter JR, Korsmeyer SJ (1993) Bcl-2-deficient mice demonstrate fulminant lymphoid apoptosis, polycystic kidneys, and hypopigmented hair. Cell 75: 229–240

Von Boehmer H (1992) Thymic selection: a matter of life and death. Immunol Today 13: 454–458

Walden PR, Eisen HN (1990) Cognate peptides induce self-destruction of CD8$^+$ cytolytic T lymphocytes. Proc Natl Acad Sci USA 87: 9015–9019

Watanabe-Fukunaga R, Brannan CI, Copeland NG, Jenkins NA, Nagata S (1992) Lympho-proliferation disorder in mice explained by defects in fas antigen that mediates apoptosis. Nature 356: 314–317

Wesselborg S, Kabelitz D (1993) Activation-driven death of human T cell clones: Time course kinetics of the induction of cell shrinkage, DNA fragmentation and cell death. Cell Immunol 148: 234–241

Wesselborg S, Janssen O, Kabelitz D (1993a) Induction of activation-driven death (apoptosis) in activated but not resting peripheral blood T cells. J Immunol 150: 4338–4345

Wesselborg S, Prüfer U, Wild M, Schraven B, Meuer SC, Kabelitz D (1993b) Triggering via the alternative CD2 pathway induces death by apoptosis in activated human T lymphocytes. Eur JImmunol 23: 2707–2710

Yonehara S, Ishii A, Yonehara M (1989) A cell-killing monoclonal antibody (anti-fas) to a cell surface antigen co-downregulated with the receptor of tumor necrosis factor. J Exp Med 169: 1747–1756

Zinkernagel RM, Hengartner H (1994) T-cell-mediated immunopathology versus direct cytolysis by virus: implications for HIV and AIDS. Immunol Today 15: 262–268

bcl-2: Antidote for Cell Death

Y. Tsujimoto

Abstract

The *bcl-2* proto-oncogene, originally identified through the study of the t(14;18) translocation present in human B-cell follicular lymphomas, is unique among oncogenes in its ability to enhance cell survival by interfering with apoptotic cell death. This finding provided the important notion to the cancer research field that inhibition of cell death might be a critical step in tumorigenesis. This idea was supported by the demonstration that several cancer genes, including oncogenes and anti-oncogenes, have activities to modulate the apoptotic process. bcl-2 exerts a death-sparing activity against apoptosis induced by a wide variety of stimuli and, therefore, appears to function at a critical step in a common process in which several different apoptotic signals converge, although the mechanism of bcl-2 function remains unknown. *bcl-2* has recently been recognized as a member of a family through the discovery of many structually related genes, some of which function like *bcl-2* while others inhibit the death-sparing function of *bcl-2* or other members. Detailed analysis of the *bcl-2* family members will provide important clues toward an understanding of the molecular basis of apoptotic cell death. Here, current information on the *bcl-2* gene and other members of this family is reviewed.

1 Introduction

In the early 1980s, the chromosomal translocations t(8;14) and t(9;22) were shown to activate the cellular oncogenes c-*myc* and c-*abl*, respectively (reviewed by Croce 1987), establishing a solid rationale for the search for new oncogenes through studies of chromosomal aberrations, namely chromosomal translocations and inversions associated with hematopoietic malignancies. The efforts in this field by several groups led to the identification of the first new oncogene *bcl-2* (*B c*ell *l*ymphoma and *l*eukemia-2) (Bakhshi et al. 1985; Clearly and Sklar 1985; Tsujimoto et al. 1985). Recently, the *bcl-2* gene was shown to have a unique biological ability to prolong cellular survival by blocking apoptotic cell death

Osaka University Medical School, Biomedical Research Center, Dept. of Medical Genetics, 2-2 Yamadaoka, Suita, Osaka 565, Japan

(Vaux et al. 1988), a major mechanism of the programmed cell death, making the *bcl-2* gene distinct among the oncogenes thus far identified.

2 Structure and Expression of the *bcl-2* Gene

The *bcl-2* gene consists of three exons and is transcribed into several species of mRNA by differential splicing, differential usage of poly(A) sites, and multiple initiation sites (Tsujimoto and Croce 1986; Seto et al. 1988). The major transcript is approximately 8 kb long and has long 5' (1.5 kb) and 3' (5.5 kb) nontranslated regions. The bcl-2 protein (26 kDa) translated from these mRNAs reveals no significant homology to known proteins, although very recently, several genes were isolated that have sequence similarity to *bcl-2* as described below. Because the splicing donor site of the second exon is within the protein coding region and because a relatively large amount of mRNA terminates within the second intron (Tsujimoto and Croce 1986; Negrini et al. 1987), a different protein product of 22 kDa (bcl-2β) is also produced, although at very low levels.

Northern analysis of adult tissues of mouse and chicken revealed that the *bcl-2* gene is expressed in a variety of tissues, with highest expression in lymphoid organs such as spleen and thymus (Negrini et al. 1987; Eguchi et al. 1992). Neuronal cells of cerebellum, cerebrum, and retina also express *bcl-2* at high levels (Eguchi et al. 1992). The *bcl-2* expression pattern in chicken embryo appears to be similar to that in adults except for some striking differences; adult muscle and bursa of *Fabricius* are negative for *bcl-2*, whereas their embryonic counterparts express high levels of the gene, providing examples of developmental regulation of *bcl-2* gene expression (Eguchi et al. 1992). Immunohistochemistry with anti-bcl-2 monoclonal antibodies has revealed a detailed pattern of *bcl-2* expression in human and mouse tissues (Hockenbery et al. 1991; LeBrun et al. 1993; Lu et al. 1993; Novack and Korsemeyer 1994). The gene is expressed in a variety of hematopoietic and non-hematopoietic tissues. It seems that bcl-2 plays a dominant role in selection on lymphocytes in the immune system and in determining the life span of neurons, lymphoid progenitors and epithelial progenitor cells in intestine and epidermis. It has also been suggested that bcl-2 functions in differentiation and morphogenesis in which cell death is an important process.

In a phylogenetic context, the *bcl-2* gene is known to be conserved down to chicken, fishes, and *Xenopus* as demonstrated by Southern blot analysis, PCR, and molecular cloning (Eguchi et al. 1992; Eguchi et al., unpubl.). No *Drosophila* homologue has been demonstrated. Recently, the *ced9* gene of *C. elegans* has been identified, which has a similar biological activity to that of the *bcl-2* and blocks the apoptotic programmed cell death of nematode cells (Hengartner et al. 1992). The *ced9* gene shows low level sequence homology with *bcl-2* and one of the exon-intron boundaries coincides with that of the *bcl-2*, suggesting that the *C. elegans ced9* gene is a *bcl-2* homologue (Hengartner and Horvitz 1994). Human *bcl-2* has been shown to block cell death occurring during development of *C. elegans* (Vaux et al. 1992).

3 Biological Function of *bcl-2*: Apoptotic Death-Sparing Activity

Since the isolation of the *bcl-2* gene as a candidate c-oncogene and the elucidation
of its structure, investigators have begun to analyze its function in lymphomagenesis. Unlike other oncogenes that are able to promote cellular proliferation,
bcl-2 has initially been shown to prolong cellular survival after depletion of IL-3 in
IL-3-dependent B and myeloid cell lines (Vaux et al. 1988). Subsequent similar
observations have been made in a variety of systems including Epstein-Barr virus-
infected lymphoblastoid cell lines (Nunez et al. 1989; Reed et al. 1989; Tsujimoto
1989), IL-3-dependent mast cell line, and IL-4-, IL-7, and GM-CSF-dependent
murine cell line (Nunez et al. 1990; Borzillo et al. 1992), and in B and T lymphocytes
isolated from *bcl-2* transgenic mice (McDonnell et al. 1989; Sentman et al. 1991;
Strasser et al. 1991a, 1992; Katsumata et al. 1992). This death-sparing capacity of
bcl-2 has been linked to an ability to block apoptosis (Hockenbery et al. 1991;
Borzillo et al. 1992). However, *bcl-2* is unable to protect against cell death induced
by IL-2 or IL-6 depletion (Nunez et al. 1990). In the case of IL-2 depletion, inability
of the *bcl-2* to prevent apoptotic death is not a function of the cell line used because
apoptosis of the same cell line induced by cytotoxic drugs is protected by *bcl-2*
(Tsujimoto, unpubl.). The fact that not all apoptosis is blocked by *bcl-2* suggests
multiple pathways of this phenomenon. As described below, the apoptosis occurring on negative selection of immature thymocytes in vivo is unaffected by the *bcl-2*
expression (Sentman et al. 1991; Strasser et al. 1991a).

4 Role of *bcl-2* in the Immune System

4.1 *bcl-2 in Lymphocyte Selection*

The topographically restricted expression of *bcl-2* in lymphoid organs (lymph
node, spleen, and thymus) and the apoptosis-blocking activity of this protein led
directly to imagination of the role of *bcl-2* in immune cell selection. Second
germinal centers in lymph node provide one of the most dramatic examples
(Pezzella et al. 1990; Hockenbery et al. 1991): immunohistochemical studies
revealed an abundance of bcl-2 in long-lived recirculating IgM$^+$/IgD$^+$B cells in the
follicular mantel of germinal centers, whereas dark zones of proliferating centro-
blasts and the basal portion of the light zone, where centrocytes undergo
apoptosis, are bcl-2-negative. B cells in the more apical portion of light zones re-
express bcl-2, and are considered to be rescued from commitment to apoptosis if,
after somatic mutation in the rearranged immunoglobulin variable region, the
immunoglobulin expressed on the cell surface shows high affinity binding to the
antigen presented on the surface of follicular dendritic cells. Although the precise
mechanism is not known, the *bcl-2* activation in the bcl-2-negative B cells of
germinal centers is achieved by the crosslinking of surface immuoglobulin (Liu
et al. 1991). This activation can also be induced by IL-1 and stimulation through
the CD40 and CD21 surface antigens (Liu et al. 1991).

As in spleen and lymph node, the thymus also displays a remarkably restricted topographical distribution of bcl-2: medullary thymocytes are bcl-2-positive, whereas most cortical thymocytes are negative, suggesting the regulation of *bcl-2* expression during T cell maturation (Pezzella et al. 1990; Hockenbery et al. 1991). The medulla is composed of $CD4^+CD8^-$ and $CD4^-CD8^+$ mature thymocytes, whereas in the cortex, $CD4^+CD8^+$ immature thymocytes predominate and most of the latter die by positive selection (selection of T cells expressing TCR that interact with the MHC complex) and negative selection (removal of T cells bearing TCR reactive with self-antigens; Blackman et al. 1990). Thus, in the immune system, *bcl-2* is expressed in lymphocytes to be required but not in lymphocytes to be removed. However, studies of *bcl-2* transgenic mice have shown that *bcl-2* is not the sole determinant of lymphocyte selection. High level expression of *bcl-2* in cortical thymocytes did not result in the rescue of auto-reactive thymocytes, indicating that the negative selection of thymocytes proceed in thymus, irrespective of *bcl-2* expression (Sentman et al. 1991; Strasser et al. 1991a). Since the escape of autoreactive T cells from negative selection is harmful, several mechanisms might operate to ensure the removal of autoreactive T cells. These findings, together with studies of variable effects of *bcl-2* on apoptosis induced by a variety of cytotoxic drugs (Tsujimoto et al., unpubl.) suggest that there are several pathways of apoptosis, some of which cannot be blocked by *bcl-2*.

4.2 Production of Autoimmune Disease

It has been shown using B cell *bcl-2* transgenic mice that the ectopic expression of the *bcl-2* gene can result in autoimmune disease. These mice often produce autoantibodies against nuclei, DNA, histones and Sm/ribonucleoprotein anti-gens (Strasser et al. 1991b). The ectopic expression of the *bcl-2* transgene in autoreactive B lymphocytes in bone marrow, where negative selection of these cells occurs, helps the cells to avoid removal, leaving the bone marrow to produce autoantibodies (Hartley et al. 1993). It is not clear whether a similar mechanism operates in a natural situation to result in autoimmune disease because in the transgenic mice, the number of autoreactive cells expressing the *bcl-2* at high levels is large so that some autoreactive B cells are likely to escape negative selection. Moreover, it is also not clear how the *bcl-2* gene is abnormally activated in autoreactive B cells in a normal situation. Viral infection is certainly one mechanism; Epstein-Barr virus infection of B cells is known to activate *bcl-2* expression through the late membrane protein (Henderson et al. 1991).

5 Role of *bcl-2* in Neuronal Tissue

The abundance of *bcl-2* mRNA in both adult and embryo neuronal tissue and the death-sparing activity of the bcl-2 protein led initially to the postulate of some role for bcl-2 in the remarkable longevity of neurons. The first experimental demonstration of bcl-2 involvement in neuronal survival came from studies using rat sympathetic neurons, whose survival in vivo and in cluture depends on the presence of nerve growth factor (NGF). The human *bcl-2* gene driven by the neuron-specific enolase promoter was microinjected into the sympathetic neurons and the effect of the presence of the human bcl-2 protein on survival after NGF depletion was studied (Garcia et al. 1992). The control cells died with in a few days, while the *bcl-2*-expressing neurons survived for a considerable period of time, providing evidence that bcl-2 exerts its death-sparing activity in neuronal cells in addition to those involved in hematopoiesis. Subsequent studies addressed similar questions in several embryonic neurons (Allsopp et al. 1993). bcl-2 was shown to protect sensory neurons that depend on NGF, brain-derived neuro-trophic factor (BDNF), and neurotrophin-3 (NT-3), but not ciliary neurotrophic factor (CNTF)-dependent ciliary neurons from cell death induced by the deple-tion of neurotrophic factors. Interestingly, exposure of the sensory neurons to CTNF renders their cell death insensitive to bcl-2 (Allsopp et al. 1993).

Transgenic mice have been generated in which neurons overproduce bcl-2 protein under the control of the neuron-specific enolase promoter. These mice appear to be useful for studies of programmed cell death of neurons during development, as well as in analyses aimed at the clinical application of bcl-2 to neurodegenerative disorders. bcl-2 overexpression has been shown to reduce naturally occurring neuronal loss during development, leading to hypertrophy of nerves including facial and optic nerves, and also inhibiting the degeneration of facial motor neurons after axotomy and neuronal death caused by permanent ischemia (Dubois-Dauphin et al. 1994; Martinou et al. 1994).

6   Role of *bcl-2* in Lymphomagenesis

The direct involvement of bcl-2 in lymphomagenesis was demonstrated in B-cell bcl-2 transgenic mice. These mice display a polyclonal follicular lymphoprolifera-tion with selectively expanded, small resting IgM^+IgD^+ mature B cells in spleen and bone marrow (McDonnell et al. 1990). After a longer time period, the mice develop aggressive B lymphoma (McDonnell and Korsmeyer 1991; Strasser et al. 1993). These observations were taken to support the model that *bcl-2* expression provides B-cell longevity, increasing the probability of secondary genetic changes that lead to neoplasia. However, the role of *bcl-2* appears to be more passive in tumorigenesis. It has recently been shown that the *c-myc* oncogene can induce apoptosis when combined with a block in cell proliferation, and that *c-myc*-induced apoptosis can be blocked by *bcl-2* (Bissonnette et al. 1992; Fanidi et al. 1992). These two oncogenes are known to cooperate in neoplastic transformation of lymphoid cells (Strasser et al. 1990; McDonnell and Korsmeyer 1991). The apoptotic cell death induced by oncogene activation is thought to represent a host defense mechanism to remove neoplastic cells and, therefore, tumorigenesis must include a step that inhibits apoptotic cell death (Fig. 1). Activation of the *bcl-2* gene with its death-sparing activity is one such mechanism. Thus, bcl-2 not only provides longevity to cells and increases the incidence of secondary genetic changes, but also reduces the risk of apoptotic death induced by the activation of a powerful oncogene such as *c-myc*. A similar scenario was observed in adenovirus transformation. Transformation by this virus requires two gene products, E1A, which, like c-myc, induces apoptosis, and E1B, which inhibits this activity (White et al. 1992). It is known that E1B is functionally similar to bcl-2. *c-myc* is involved in a variety of tumors, however, it is unclear whether a functional equivalent of *bcl-2* always exists.

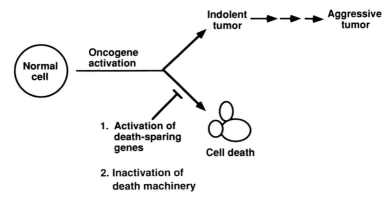

Fig. 1. Inhibition of cell death in tumorigenesis. A normal cell becomes neoplastic, initially indolent and subsequently more aggressive by accumulating genetic alterations. Activation of some oncogenes induces apoptosis, which is thought to be a host defense mechanism to remove neoplastic cells. Activation of *bcl-2* gene is one of the mechanisms which inhibit the cell death process, leading cells toward a neoplastic state

7 Role of *bcl-2* in Other Systems

Accumulating information points the role of bcl-2 in regulating cell death in many biological systems.

7.1 Virus Persistent Infection

Sindbis virus, a single-stranded RNA virus, induces lytic infection in the most vertebrate cell lines, killing cells by apoptosis, but persistently infects postmitotic neurons. Since bcl-2 blocks neuronal cell death, the possible relationship between bcl-2 and persistent productive viral infection was addressed (Levine et al. 1993). Sindbis virus infection of rat prostatic adenocarcinoma cells resulted in apoptotic death, whereas viral infection of *bcl-2*-transfected cells resulted in long term, persistent productive infection. Thus, bcl-2 appears to play a role in determining the fate of virus-infected cells, regarding lytic or persistent infection.

7.2   Epithelial Cells

Epithelial cells in many organs express bcl-2, although in some cases, expression is topographically restricted. For example, only regenerative basal crypt cells in the intestine are highly bcl-2-positive (Hockenbery et al. 1991), suggesting that bcl-2 may serve to maintain the stem cell pool by allowing certain cells to avoid senescence. Another striking example is the expression of bcl-2 in uterus epithelium cells during the menstrual cycle, which is regulated by hormones such as estrogen and progesterone. These ephithelial cells progress in three steps: proliferative, secretory, and menstrual phases. bcl-2 expression is restricted to the proliferative and early secretory phases, disappearing before the menstrual phase (Ohtsuki et al. 1994). The massive cell death that characterizes menstrual phase occurs after the bcl-2 protein disappears, suggesting that the disappearance of bcl-2 might facilitate cell death or induce cell differentiation.

7.3   Morphogenesis

Since programmed cell death is an important process in morphogenesis, bcl-2 most likely participates in this process. Based on analysis of bcl-2 gene expression in human fetal tissues, LeBrun et al. (1993) described that bcl-2 is present in regions characterized by inductive interaction, suggesting a role for bcl-2 in morphogenesis. However, thus far, experimental evidence for such a role is lacking. Transgenic animals in which the bcl-2 gene is expressed in regions that include cell population being removed during morphogenesis, for example, limb buds, representing a promising system to address morphological disturbance by ectopic bcl-2 expression.

8   Analysis of bcl-2-Deficient Mice

Gene targeting in embryonic stem cells provides a powerful tool to generate animals with loss of gene function (Mansour et al. 1988). Since bcl-2 is widely expressed in embryonic and adult tissues, development of bcl-2 deficient mice derived through this targeting method (Veis et al. 1993; Nakayama et al. 1994; Kamada et al. 1995) provides a useful system to analyze bcl-2 function in vivo. bcl-2 deficiency does not interfere with embryonic development, but abnormalities become evident soon after birth, although the severity of abnormalities varies among bcl-2-/-siblings. bcl-2-/-mice generally fall into two groups; one displays severe growth retardation and dies in several weeks, and the other shows moderate growth retardation and survives much longer. These mice generally display small ears relative to their body size. Most striking anomalies are seen in kidney which becomes polycystic. The kidney abnormalities appear to begin as early as the embryonic stage (Y.Tsujimoto, unpubl.). The size of the thymus and spleen, and of the lymphocyte subclass populations in each organ, is normal in young and healthy *bcl-2-/-*mice. In contrast, ill mice show accelerated apoptotic cell death of lymphocytes in the lymphoid organs, resulting in atrophic spleen and thymus. Thus, it appears that *bcl-2* is not essential for development and maturation of lymphocytes, but lack of *bcl-2* results in increased susceptibility of lymphocytes to cell death. Accelerated apoptotic death of *bcl-2-/-*lymphocytes was also demonstrated in *bcl-2-/-*mutant chimeric mice (Nakayama et al. 1993). *bcl-2-/-*mice from the group surviving longer turn gray in the second hair follicle cycle, apparently due to loss of melanocytes (Y.Tsujimoto, unpubl.), although the hyperoxidation of melanin caused by *bcl-2*-deficiency has been proposed to underlie the hypopigmentation (Veis et al. 1993). Careful examination of our *bcl-2-/-*mice has revealed additional abnormalities in several organs such as the small intestine, where villi are short and bizzare in shape (Kamada et al. 1995). In the normal small intestine, the epithelial layer consists of several different types of cells, all of which derive from progenitor cells at the crypt. Daughter cells of a small number of progenitors with long-term proliferative ability divide rapidly and differentiate as they move either upward or downward. The shortness and bizarre shape of villi in *bcl-2-/-*mice may reflect the decrease in dividing cells in the crypts and accelerated exfoliation of epithelial cells that was not restricted at the villus tip. Since a high level of *bcl-2* expression has been reported in the crypts (Hockenbery et al. 1991), the decrease in dividing cells there may result directly from the lack of bcl-2 protein. However, despite the abnormal architecture of villi and crypts, *bcl-2-/-*mice retain absorptive epithelial, goblet and Paneth's cells in the small intestine, suggesting that the bcl-2 protein may not be essential for differentiation of the progenitor cells. The fewer mitotic figures seen in small intestines of bcl-2-/- mice than in control mice are likely the result of accelerated cell death of the progenitors in the absence of *bcl-2*, suggesting an essential role for bcl-2 in the maintenance of long-lived crypt progenitors. Together, analyses of *bcl-2*-deficient mice suggest a dominant role for bcl-2 protein in vivo in maintaining the appropriate life span of stem cells as well as differentiated cells. Although such analyses have not directly indicated the involvement of *bcl-2* in positive selection of lymphocytes, such a role cannot be excluded since *bcl-2* deficiency might be compensated by other genes with similar activity. Compensation of *bcl-2* deficiency by other genes might also explain the absence of detectable abnormalities in some organs that normally express high levels of *bcl-2*, such as neuronal tissues, including brain.

9   Subcellular Localization of *bcl-2*: Multiple Membrane Locations

Availability of antibodies against bcl-2 was key in showing that bcl-2 is a membrane protein (Tsujimoto et al. 1987). Subsequent efforts by several investigators to elucidate the subcellular localization of the bcl-2 protein as a step toward understanding its biochemical function revealed controversial results. In 1989, immunofluoresence-microscopy was used to localize bcl-2 in plasma membrane and perinuclear endoplasmic reticulum membrane (Chen-Levy et al. 1989), and in the next year, biochemical fractionation procedures and laser-scanning confocal immunomicroscopy indicated that the bcl-2 protein localizes exclusively in mitochondrial inner membrane (Hockenbery et al. 1990). This suggested that mitochondria is involved in apoptosis and that bcl-2 exerts its role through mitochondrial function. However, recently, several reports suggest that the bcl-2 protein localizes in other membrane compartments such as endoplasmic reticulum membrane and nuclear membrane as well as mitochondria, using immunofluoresence microscopy (Alnemri et al. 1992; Jacobson et al. 1993). Our group and others used the immunoelectronmicroscopy for more detailed localization of *bcl-2* (Monaghan et al. 1992; Krajewski et al. 1993; Akao et al. 1995), detecting in multiple membrane sites including nuclear outer membrane, rough and smooth endoplasmic reticulum membrane, and mitochondrial membranes. The presence of *bcl-2* in nuclear outer membrane was also confirmed by biochemical fractionation (Akao et al. 1994). This multisite membrane distribution of bcl-2 suggests an important role for this protein in several different membrane compartments.

The C-terminal portion of bcl-2 which consists of a stretch of hydrophobic amino acid residues, has been implicated in membrane localization: deletion of this stretch results in translocation of the bcl-2 protein from the membrane compartments to the cytosol (Chen-Levy and Cleary 1990; Tsujimoto et al., unpubl.). Because the same deletion also abolishes the death-sparing activity of bcl-2 (Tsujimoto et al., unpubl.), the membrane localization of bcl-2 might be crucial for its biological activity although the species conservation of the C-terminal portion suggests an important function for this region (Eguchi et al. 1992).

10   Biochemical Function of bcl-2 Protein

Two models have been proposed for the biochemical role of the bcl-2 protein. One model suggests that the protein regulates intracellular levels of calcium ion which have been implicated in cell death (Baffy et al. 1993; Lam et al. 1994), whereas the other model proposes that bcl-2 protein regulates production and/or activity of reactive oxygen species (ROS) (Hockenbery et al. 1993; Kane et al. 1993), which are thought to be common mediators of apoptotic cell death (Buttke and Sandstrom 1994). Although the model of *bcl-2* function against ROS has been widely accepted, we (Shimizu et al. 1995) and others (Jacobson and Raff 1995) have recently shown that *bcl-2* can protect against cell death induced under hypoxic conditions where essentially no ROS are formed, leading us to conclude that *bcl-2* exerts its cell death-sparing effect by a mechanism independent of ROS. Analyses also strongly suggest that ROS is just one of many inducers of apoptosis. The fact that *bcl-2* acts in a protective manner in almost all instances of apoptosis suggests that it functions at a critical step of a common pathway into which apoptotic signals triggered by a variety of stimuli converge (Fig. 2). Cystein proteases such as ICE have been suggested to represent a common mediator of apoptosis (Yuan et al. 1993; Gagliardini et al. 1994) and it is possible that *bcl-2* acts in the close vicinity of such proteases (Fig. 2).

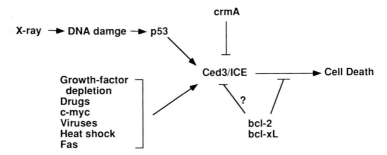

Fig. 2. Framework of apoptosis. Apoptotic signals induced by a variety of stimuli converge into a common pathway which is inhibited by *bcl-2*. Cystein proteases such as ICE, the mammalian homologue of *C. elegans* death gene *ced-3* (Yuan et al. 1993), is a candidate for a common mediator of apoptosis. The cowpox virus gene *crmA* encodes a 38-kDa protein that can specifically inhibit ICE activity (Ray et al. 1992)

11 *bcl-2* Related Genes and bcl-2 Associated Proteins

Recent investigations have clearly shown that *bcl-2* is just one member of a gene family. This family includes *bax*, *bcl-x*, *A1*, *bad* and *bak*, all of which are similar in size (20 to 26 kDa). bax was identified as protein that interacts with the bcl-2 protein (Oltvai et al. 1993). Sequence analysis of *bax* revealed homology to *bcl-2* at two limited regions, called BH1 and BH2. The bax protein can form a homodimer, and by heterodimerizing with bcl-2, abolishes the death-sparing activity of bcl-2. The *bcl-x* gene was isolated by screening libraries with *bcl-2* probes under less stringent conditions (Boise et al. 1993). *bcl-x* encodes two proteins, bcl-xL (large) and bcl-xS (small), by differential splicing. bcl-xL is as potent as bcl-2 in inhibiting cell death induced by various stimuli including growth factor withdrawal (Boise et al. 1993) and hypoxia (Shimizu et al. 1995). By contrast, bcl-xS inhibits the ability of bcl-2 to enhance the survival of growth factor-deprived cells. bcl-xL is a major component of *bcl-x* gene products, whereas bcl-xS appears to be present at very low levels or not at all (Boise et al. 1993; Gonzalez-Garcia et al. 1994). The *bad* gene was identified as a gene encoding a bcl-2 binding protein (Yang et al. 1995). The bad protein hetero-dimerizes more strongly with bcl-xL than with bcl-2, and inhibits the death-sparing activity of bcl-xL without any prominent effect on bcl-2 activity (Yang et al. 1995). The recently identified *bak* gene, another member of the *bcl-2* family, was shown to have a similar function to that of *bax* (Chittenden et al. 1995; Farrow et al. 1995; Kiefer et al. 1995). It is not yet clear whether *bax* and *bak* act to mitigate the protective function of bcl-2 or bcl-xL, or to activate the apoptotic pathway that is inhibited by bcl-2 or bcl-xL. The genes, *mcl-1* which encodes a 37-kDa protein and has a region of homology with *bcl-2* (Kozopas et al. 1993), and *A1* (Lin et al. 1993) are induced upon myeloid cell differentiation and upon GM-CSF stimulation, respectively; mcl-1 appears to suppress apoptosis since *mcl-1* overexpression inhibits *myc*-induced apoptosis (Reynolds et al. 1994), whereas a cell death-modulating activity of A1 has not yet been demonstrated. A positive interaction among members of the *bcl-2* family, although not in all combinations, has been demonstrated using a yeast two-hybrid gene assay (Sato et al. 1994). In addition, Epstein-Barr virus and African swine fever virus have been shown to carry a gene, *BHRF1* (Pearson et al. 1987) and *LMW5-HL* (Neilan et al. 1993), respectively, with limited sequence similarity to the *bcl-2* gene; thus far, only *BHRF1* has been shown to exert death-sparing activity. The *E1B* gene of adenovirus is also known to have death-sparing activity and homology to *bcl-2* at a very limited region. The use of a yeast two hybrid system or a screening bacterial expression library has identified several genes (*R-ras-23*, *nip-1*, *nip-2*, *nip-3* and *bag-1*) whose protein products interact with the bcl-2 protein but do not share any sequence similarity with the *bcl-2* family genes. A ras-related protein, R-ras p23 with GTP-binding activity has been shown to interact with the bcl-2 protein (Fernandez-Sarabia and Bischoff 1993), but the function of R-ras p23 remains unknown. *Nip-1, 2* and *3*, isolated as genes whose products interact with the E1B 19-kDa protein (Boyd et al. 1994), have been shown to bind the bcl-2 protein, although their biological activity also remains unknown. *bag-1* encodes a protein of 219 amino acid residues and enhances the protective activity of bcl-2 against Fas-and staurosporin-induced cell death (Takayama et al. 1995).

12 Epilogue

Apoptosis is a theme not only of general biology, but also of medical science because abnormal activation or inactivation of apoptosis is thought to be one of the underlying elements of diseases such as cancer and neurodegenerative disorders. Apoptosis can be induced by a variety of stimuli whose signals most likely converge into a common pathway. An understanding of the molecular basis of apoptosis as well as some insights into treatment or prevention of diseases awaits elucidation of the critical common mediators of apoptosis. In this context, analyses of functions of *bcl-2* family members will provide important clues about the molecular mechanism of apoptosis, since bcl-2 is able to protect nearly all forms of apoptosis and therefore probably exerts its death-sparing activity by functioning against a common mediator of apoptosis.

References

Akao Y, Ohtsuki Y, Kataoka S, Ito Y, Tsujimoto Y (1994) Multiple subcellular localization of bcl-2: detection in nuclear outer membrane, endoplasmic reticulum and mitochondrial inner and outer membranes. Cancer Res 54: 2468–2471
Allsopp TE, Wyatt S, Paterson HF, Davies AM (1993) The proto-oncogene *bcl-2* can selectively rescue neurotrophic factor-dependent neurons from apoptosis. Cell 73: 295–307
Alnemri ES, Robertson NM, Fernandes TF, Croce CM, Litwack G (1992) Overexpressed full-length human *bcl-2* extends the survival of baculovirus-infected Sf9 insect cells. Proc Natl Acad Sci USA 89: 7295–7299

Baffy G, Miyashita T, Williamson JR, Reed JC (1993) Apoptosis induced by withdrawal of interleukin-3 (IL-3) from an IL-3-dependent hematopoietic cell line is associated with re-partitioning of intracellular calcium and is blocked by enforced *bcl-2* oncoprotein production. J Biol Chem 268: 6511–6519

Bakhshi A, Jensen JP, Goldman P, Wright JJ, McBride OW, Epstein AL, Korsmeyer SJ (1985) Cloning the chromosomal breakpoint of t(14; 18) human lymphoma: clustering around J_H on chromosome 14 and near a transcriptional unit on 18. Cell 41: 889–906

Bissonnette RP, Echeverri F, Mahboubi A, Green DR (1992) Apoptotic cell death induced by c-*myc* is inhibited by *bcl-2*. Nature 359: 552–554

Blackman M, Kappler J, Marrack P (1990) The role of the T cell receptor in positive and negative selection of developing T cells. Science 248: 1335–1338

Boise LH, Gonzalez-Garcia M, Postema CE, Ding L, Lindsten T, Turka LA, Mao X, Nunez G, Thompson CB (1993) *bcl-x,* a *bcl-2* related gene that functions as a dominant regulator of apoptotic cell death. Cell 74: 597–608

Borzillo GV, Endo K, Tsujimoto Y (1992) Bcl-2 confers growth and survival advantage to IL-7-dependent early pre-B cells which become factor-independent by a multiple process in culture. Oncogene 7: 869 – 876

Boyd JM, Malstrom S, Subramanian T, Venkatesh LK, Schaeper U, Elangovan B, D'Sa-Eipper C, Chinnadurai G (1994) Adenovirus E1B 19kDa and bcl-2 proteins interact with a common set of cellular proteins. Cell 79: 341–351

Buttke TM, Sandstrom PA (1994) Oxidative stress as a mediator of apoptosis. Immunol Today 7: 7–10

Chen-Levy Z, Cleary ML (1990) Membrane topology of the *bcl-2* proto-oncogene protein demonstrated in vitro. J Biol Chem 265: 4929–4933

Chen-Levy Z, Nourse J, Cleary ML (1989) The *bcl-2* candidate proto-oncogene product is a 24-kilodalton integral membrane protein highly expressed in lymphoid cell lines and lymphoma carrying the t(14; 18) translocation. Mol Cell Biol 9: 701–710

Chittenden T, Harrington EA, O'Connor R, Flemington C, Lutz RJ, Evan GI, Guild BC (1995) Induction of apoptosis by the *bcl-2* homologue *bak*. Nature 374: 733–736

Cleary ML, Sklar J (1985) Nucleotide sequence of at (14; 8) chromosomal breakpoint in follicular lymphoma and demonstration of a breakpoint cluster region near a transcriptionally active locus on chromosome 18. Proc Natl Acad Sci USA 81: 593–597

Croce CM (1987) Role of chromosome translocations in human neoplasia. Cell 49: 155–156

Dubois-Dauphin M, Frankowski H, Tsujimoto Y, Huarte J, Martinou J-C (1994) Neonatal motoneurons overexpressing the *bcl-2* protooncogene in transgenic mice are protected from axotomy-induced cell death. Proc Natl Acad Sci USA 91: 3309–3313

Eguchi Y, Ewert DL, Tsujimoto Y (1992) Isolation and characterization of the chicken *bcl-2* gene: expression in a variety of tissues including lymphoid and neuronal organs in adult and embryo. Nucl Acid Res 20: 4187–4192

Fanvini A, Harrington EA, Evan GI (1992) Cooperative interaction between c-*myc* and *bcl-2* proto-oncogenes. Nature 359: 554–556

Farrow SN, White JH, Martinou I, Raven T, Pun K-T, Grinham CJ, Martinou J-C, Brown R (1995) Cloning of a *bcl-2* homologue by interaction with adenovirus E1B 19 K. Nature 374: 731–733

Fernandez-Sarabia MJ, Bischoff JR (1993) *bcl-2* associates with the *ras*-related protein *R-ras* p23. Nature 366: 274–275

Gagliardini V, Fernandez P-A, Lee RKK, Drexler CA, Rotello RJ, Fishman MC, Yuan J (1994) Prevention of vertebrate neuronal death by the *crmA* gene. Science 263: 826–828

Garcia I, Martinou I, Tsujimoto Y, Martinou J-C (1992) Prevention of programmed cell death of sympathetic neurons by the *bcl-2* proto-oncogene. Science 258: 302–304

Gonzalez-Garcia M, Perez-Ballestero R, Ding L, Duan L, Boise LH, Thompson CB, Nunez G (1994) bcl-xL is the major bcl-x mRNA form expressed during murine development and its product localizes to mitochondria. Development 120: 3033–3042

Hartley SB, Cooke MP, Fulcher DA, Harris AW, Cory S, Basten A, Goodnow CC (1993) Elimination of self-reactive B lymphocytes proceeds in two stages: arrested development and cell death. Cell 72: 325–335

Henderson S, Rowe M, Gregory C, Croom-Carter D, Wang F, Iongnecker R, Kieff E, Rickinson D (1991) Induction of *bcl-2* expression by Epstein-Barr virus latent membrane protein 1 protects infected B cells from programmed cell death. Cell 65: 1107–1115

Hengartner MO, Ellis R, Horvitz R (1992) Caenorhabditid elegans gene *ced-9* protects cells from programmed cell death. Nature 356: 494–499

Hengartner MO, Hortvitz HR (1994) C. elegans cell survival gene ced-9 encodes a functional homolog of the mammalian proto-oncogene bcl-2. Cell 70: 665–676

Hockenbery D, Nunez G, Milliman C, Schreber RD, Korsmeyer SJ (1990) *bcl-2* is an inner mitochondrial membrane protein that blocks programmed cell death. Nature 348: 334–336

Hockenbery DM, Zutter M, Hickey W, Moon N, Korsmeyer SJ (1991) *bcl-2* protein is topographically restricted in tissues characterized by apoptotic cell death. Proc Natl Acad Sci USA 88: 6961–6965

Hockenbery DM, Oltvai ZN, Yin X-M, Milliman CL, Korsmeyer SJ (1993) *bcl-2* functions in an antioxidant pathway to prevent apoptosis. Cell 75: 241–251

Jacobson MD, Burne JF, King MP, Miyashita T, Reed JC, Raff MC (1993) *bcl-2* blocks apoptosis in cells lacking mitochondrial DNA. Nature 361: 365–369

Jacobson MD, Raff MC (1995) Programmed cell death and *bcl-2* protection in very low oxygen. Nature 374: 814–816

Kamada S, Shimono A, Shinto Y, Tsujimura T, Takahashi T, Noda T, Kitamura Y, Kondoh H, Tsujimoto Y (1995) *Bcl-2* deficiency in mice leads to pleiotropic abnormalities: accelerated lymphoid cell death in thymus and spleen, polycystic kidney, hair hypopigmentation and distorted small intestine. Cancer Res 55: 354–359

Kane DJ, Sarafian TA, Anton R, Hahn H, Gralla EB, Valentine JS, Ord T, Bredesen DE (1993) *bcl-2* inhibition of neural death: decreased generation of reactive oxygen species. Science 262: 1274–1277

Katsumata M, Siegel RM, Louie DC, Miyashita T, Tsujimoto Y, Nowell PC, Green MI, Reed JC (1992) Differential effects of *bcl-2* on T and B cells in transgenic mice. Proc Natl Acad Sci USA 89: 11376–11380

Kiefer MC, Brauer MJ, Powers VC, Wu JJ, Umansky SR, Tomei LD, Barr PJ (1995) Modulation of apoptosis by the widely distributed *bcl-2* homologue *bak*. Nature 374: 736–739

Kozopas KM, Yang T, Buchan HL, Zhou P, Craig RW (1993) *Mcl-1*, a gene expressed in programmed myeloid cell differentiation, has sequence similarity to *bcl-2*. Proc Natl Acad Sci USA 90: 3516–3520

Krajewski S, Tanaka S, Takayama S, Schibler MJ, Fenton W, Reed JC (1993) Investigation of the subcellular distribution of the *bcl-2* oncoprotein: residence in the nuclear envelope, endoplasmic reticulum, and outer mitochondrial membranes. Cancer Res 53: 4701–4714

Lam M, Dubyak G, Chen L, Nunez G, Miesfeld RL, Distelhost CW (1994) Evidence that *bcl-2* represses apoptosis by regulating endoplasmic reticulum-associated Ca^{2+} fluxes. Proc Natl Acad Sci USA 91: 6569–6573

Levine B, Huang Q, Isaacs JT, Reed JC, Griffin DE, Hardwick JM (1993) Conversion of lytic to persistent alphavirus infection by the *bcl-2* cellular oncogene. Nature 361: 739–742

LeBrun DP, Warnke RA, Cleary ML (1993) Expression of *bcl-2* in fetal tissues suggests a role in morphogenesis. Am J Pathol 142: 743–753

Lin EY, Orlofsky A, Berger MS, Prystowsky MB (1993) Characterization of A1, a novel hemopoietic-specific early-response gene with sequence similarity to *bcl-2*. J Immunol 151: 1979–1988

Liu Y-J, Mason DY, Johnson GD, Abbot S, Gregory CD, Hardie DL, Gordon J, MacLennan ICM (1991) Germinal center cells express bcl-2 protein after activation by signals which prevent their entry into apoptosis. Eur J Immunol 21: 1905–1910

Lu Q-L, Poulsom R, Wong L, Hanby AM (1993) *Bcl-2* expression in adult and embryonic non-haematopoietic tissues. J Pathol 169: 431–437

Mansour S, Thomas KR, Capecchi MR (1988) Disruption of the proto-oncogene int-2 in mouse embryo-derived stem cells: a general strategy for targeting mutations to non-selectable genes. Nature 336: 348–352

Martinou J-C, Dubois-Dauohin M, Staple JK, Rodriguez I, Frankowski H, Missotten M, Albertini P, Talabot D, Catsicas S, Pietra C, Huarte J (1994) Overexpression of *bcl-2* in transgenic mice protects neurons from naturally occurring cell death and experimental ischemia. Neuron 13: 1017–1030

McDonnell TJ, Deane N, Platt FM, Nunez G, Jaeger U, McKearn JP, Korsmeyer SJ (1989) *bcl-2*-immunoglobulin transgenic mice demonstrate extended B cell survival and follicular lympho-proliferation. Cell 57: 79–88

McDonnell TJ, Korsmeyer SJ (1991) Progression from lympoid hyperplasia to high-grade malignant lymphoma in mice transgenic for the t(14; 18). Nature 349: 254–256

Monaghan P, Robertson D, Amos TAS, Dyer MJS, Mason DY, Greaves MF (1992) Ultrastructural localization of bcl-2 protein. J Histochem Cytochem 40: 1819–1825

Nakayama K-I, Nakayama K, Negishi I, Kuida K, Shinkai Y, Louie MC, Fields LE, Lucas PJ, Stewart V, Alt FW, Loh DY (1993) Disappearance of the lymphoid system in *bcl-2* homozygous mutant chimeric mice. Science 261: 1584–1587

Nakayama K, Nakayama K-I, Negishi I, Kuida K, Sawa H, Loh DY (1994) Targeted disruption of *bcl-2αβ* in mice: occurrence of gray hair, polycystic kidney disease, and lymphocytopenia. Proc Natl Acad Sci USA 91: 3700–3704

Neilan JG, Lu Z, Afonso CL, Kutish GF, Sussman MD, Rock DL (1993) An African swine fever virus gene with similarity to the protooncogene *bcl-2* and the Epstein-Barr virus gene BHRF1. J Virol 67: 4391–4394

Negrini M, Silini E, Kozak C, Tsujimoto Y, Croce CM (1987) Molecular analysis of *mbcl-2*: structure and expression of the murine gene homologous to the human gene involved in follicular lymphoma. Cell 49: 455–463

Novack DV, Korsemeyer SJ (1994) Bcl-2 protein expression during murine development. Am J Pathol 145: 61–73

Nunez G, Seto M, Seremetis S, Ferrero D, Grignani F, Korsmeyer SJ, Dalla-Favera R (1989) Growth- and tumor-promoting effects of deregulated *bcl-2* in human B-lymphoblastoid cells. Proc Natl Acad Sci USA 86: 4589–4593

Nunez G, London L, Hockenbery D, Alexander M, McKearn JP, Korsmeyer SJ (1990) Deregulated *bcl-2* gene expression selectively prolongs survival of growth factor-deprived hematopoietic cell lines. J Immunol 144: 3602–3610

Oltvai ZN, Milliman CL, Korsmeyer SJ (1993) bcl-2 heterodimerizes in vivo with a conserved homolog, bax, that accelerates programmed cell death. Cell 74: 609–619

Pearson GR, Luka J, Petti L, Sample M, Birkenback M, Braun D, Kieff E (1987) Identification of an Epstein-Barr virus early gene encoding a second component of the restricted early antigen complex. Virology 160: 151–161

Pezzella F, Tse AGD, Cordell JL, Pulford KAF, Gatter KC, Mason DY (1990) Expression of the *bcl-2* oncogene protein is not specific for the 14;18 chromosomal translocation. Am J Pathol 137: 225–232

Ray CA, Black RA, Kronheim SR, Greenstreet TA, Sleath PR, Salvesen GS, Pickup DJ (1992) Viral inhibition of inflammation: cowpox virus encodes an inhibitor of the interleukin-1β converting enzyme. Cell 69: 597–604

Reed JC, Haldar S, Cuddy MP, Croce CM, Makover D (1989) Deregulated *bcl-2* expression enhances growth of a human B cell line. Oncogene 4: 1123–1127

Reynolds JE, Yang T, Qian L, Jenkinson JD, Zhou P, Eastman A, Craig RW (1994) Mcl-1, a member of the *bcl-2* family, delays apoptosis by c-myc overexpression in Chinese hamster ovary cells. Cancer Res 54: 6348–6352

Sato T, Hanada M, Bodrug S, Irie S, Iwama N, Boise LH, Thompson CB, Golemis E, Fong L, Wang H-G, Reed JC (1994) Interaction among members of the bcl-2 protein family analyzed with a yeast two-hybrid system. Proc Natl Acad Sci USA 91: 9238–9242

Sentman CL, Shutter JR, Hockenbery D, Kanagawa O, Korsmeyer SJ (1991) bcl-2 inhibits multiple forms of apoptosis but not negative selection of thymocytes. Cell 67: 879–888

Seto M, Jaeger U, Hockett RD, Graninger W, Bennett S, Goldman P, Korsmeyer SJ (1988) Alternative promoters and exons, somatic mutations and deregulation of the *bcl-2*-Ig fusion gene in lymphoma. EMBO J 7: 123–131

Shimizu S, Eguchi Y, Kosaka H, Kamiike W, Matsuda H, Tsujimoto Y (1995) Prevention of hypoxia-induced cell death by bcl-2 and bcl-xL. Nature 374: 811–813

Strasser A, Harris AW, Bath M, Cory S (1990) Novel primitive lymphoid tumors induced in transgenic mice by cooperation between *myc* and *bcl-2*. Nature 348: 331–333

Strasser A, Harris AW, Cory S (1991a) *bcl-2* transgene inhibits T cell death and perturbs thymic self-censorship. Cell 67: 889–899

Strasser A, Whittingham S, Vaux DL, Bath ML, Adams JM Cory S, Harris A (1991b) Enforced *bcl-2* expression in B-lymphoid cells prolongs antibody responses and elicits autoimmune disease. Proc Natl Acad Sci USA 88: 8661–8665

Strasser A, Harris AW, Vaux DL, Webb E, Bath ML, Elefanty AG, Adams JM, Cory S (1992) The role of *bcl-2* in lymphoid differentiation and neoplastic transformation. Curr Top Microbiol Immunol 182: 299–302

Strasser A, Harris AW, Cory S (1993) Eµ-*bcl-2* transgene faciliates spontaneous transformation of early pre-B and immunoglobulin-secreting cells but not T cells. Oncogene 8: 1–9

Takayama S, Sato T, Krajewski S, Kochel K, Irie S, Millan JA, Reed JC (1995) Cloning and functional analysis of bag-1: a novel bcl-2-binding protein with anti-cell death activity. Cell 80: 279–284

Tsujimoto Y (1989) Overexpression of the human *bcl-2* gene product results in growth enhancement of Epstein-Barr virus-immortalized B cells. Proc Natl Acad Sci USA 86: 1958–1962

Tsujimoto Y, Croce CM (1986) Analysis of the structure, transcripts and protein products of *bcl-2*, the gene involved in human follicular lymphoma. Proc Natl Acad Sci USA 83: 5214–5218

Tsujimoto Y, Cossman J, Jaffe E, Croce CM (1985) Involvement of the *bcl-2* gene in human follicular lymphoma. Science 228: 1440–1443

Tsujimoto Y, Ikegaki N, Croce CM (1987) Characterization of the protein product of *bcl-2*, the gene involved in human follicular lymphoma. Oncogene 2: 3–7

Vaux DL, Cory S, Adams JM (1988) *bcl-2* gene promotes haemopoietic cell surval and cooperates with c-*myc* to immortalize pre-B cells. Nature 335: 440–442

Vaux DL, Weissman IL, Kim SK (1992) Prevention of programmed cell death in *Caenorhabditis elegans* by human *bcl-2*. Science 258: 1955–1957

Veis DJ, Sorenson CM, Shutter JR, Korsmeyer SJ (1993) *Bcl-2*-deficient mice demonstrate fulminant lymphoid apoptosis, polycystic kidneys, and hypopigmented hair. Cell 75: 229–240

White E, Sabbatini P, Debba M, Wold WSM, Kusher DI, Gooding LR (1992) The 19-kilodalton adenovirus E1B transforming protein inhibits programmed cell death and prevents cytolysis by tumor necrosis factor α. Mol Cell Biol 12: 2570–2580

Yang E, Zha J, Jockel J, Boise LH, Thompson CB, Korsmeyer SJ (1995) Bad, a heterodimeric partner for bcl-xL and bcl-2, displaces bax and promotes cell death. Cell 80: 285–291

Yuan J, Shaham S, Ledoux S, Ellis HM, Horvitz HR (1993) The *C. elegans* cell death gene *ced-3* encodes a protein similar to mammalian interleukin-1β-converting enzyme. Cell 75: 641–652

Apoptosis Mediated by the Fas System

S. Nagata

Abstract

Fas is a cell-surface protein mediating apoptosis. Fas ligand is a type II membrane protein and induces apoptosis by binding to Fas. Fas ligand is expressed in activated T cells, and works as an effector of cytotoxic lymphocytes. Molecular and gnetic analysis of Fas and Fas ligand indicated that mouse lymphoproliferation mutation (*lpr*) and generalized lymphoproliferative disease (*gld*) are mutations of Fas and Fas ligand, respectively. The *lpr* of *gld* mice develop lymphadenopathy, and suffer from autoimmune disease. Based on these phenotypes and other studies, it was concluded that the Fas system is involved in the apoptotic process during T-cell development, specifically peripheral clonal deletion or activation-induced suicide of mature T cells. In addition to the activated lymphocytes, Fas is expressed in the liver, heart, and lung.

Administration of agonistic anti-Fas antibody into mice induced apoptosis in the liver and quickly killed the mice, causing liver damage. These findings suggest that the Fas system plays a role not only in the physiological process of lymphocyte development, but also in the cytotoxic T-lymphocyte-mediated disease such as fulminant hepatitis.

1 Introduction

Mammalian development and homeostasis are tightly regulated not only by proliferation and differentiation of cells, but also by cell death or apoptosis (Ellis et al. 1991; Raff 1992). For example, many cells die by apoptosis during embryogenesis, metamorphosis, endocrine-dependent tissue atrophy, and normal tissue turnover (Wyllie et al. 1980; Walker et al. 1988). Proliferation and differentiation of cells are regulated by cytokines and their receptors (Arai et al. 1990). More than dozens of cytokines and their receptors are known to be involved in proliferation and differentiation of cells. Cytokines produced by various effector cells such as activated T lymphocytes or macrophages bind to their receptor in the target cells, and activates the secondary signaling pathway such as kinases,

Osaka Bioscience Institute, 6-2-4 Furuedai, Suita, Osaka 565, Japan adenylate cyclase, and Ca^{2+} mobilization. These signalings are finally incarnated as induction of specific gene expression for proliferation and differentiation.

During apoptosis, microvilli of plasma membrane disappear, nuclei are condensed and fragmented, and chromosomal DNA is extensively degraded into units of nucleosome (Raff 1992). Currently, the kinds of signaling molecules involved in apoptosis are unknown except for several genes in *Caenorhabditis elegans* (Ellis et al. 1991) and some oncogenes in mammalian systems. The Fas antigen (Fas) is a cell-surface protein belonging to the tumor necrosis factor (TNF)/nerve growth factor (NGF) receptor family. Our recent studies on Fas and its ligand indicated that Fas ligand and Fas work as a death factor and its receptor, and suggest that in some cases, apoptosis is also regulated by cytokines and their receptors. Here, I present the summary of the Fas/Fas ligand system elucidated in my laboratory over 5 years, and discuss its physiological roles.

2   Fas, a Receptor for a Death Factor

2.1   Molecular Properties of Fas

In 1989, Yonehara et al. (1989) established several mouse monoclonal antibodies against human FS-7 cells. One of the antibodies, called the anti-Fas (FS-7 associated surface antigen) antibody, had a cytolytic activity against cells expressing the Fas antigen. In the same year, Trauth et al. (1989) also established a similar monoclonal antibody against human cell surface protein. The antibody was designated anti-APO1 antibody because it caused apoptosis of cells expressing the antigen. There were two possibilities concerning the function of the Fas and APO1 antigens. The antigen can be the receptor for a molecule that has cytolytic activity, and the monoclonal antibodies work as an agonist or mimic the effect of the ligand. The other possibility is that the antigen can be the receptor for a growth factor or a molecule essential for growth-signal transduction. In this case, the antibody works as an antagonist, and it blocks the binding of the growth factor to inhibit growth signal transduction.

To test these possibilities, we isolated human Fas cDNA from human lymphoma KT-3 cell line by expression cloning (Itoh et al. 1991). The amino acid sequence predicted from the cDNA indicated that Fas consists of 325 amino acids with a signal sequence at the N-terminus and a transmembrane domain in the middle of the molecule. Later, Oehm et al. (1992) purified the APO1 antigen as a protein of 45 kDa from the membrane fraction of human lymphoblastoid cell line. Molecular cloning of the APO1 antigen cDNA using the amino acid sequence information indicated that the APO1 antigen is identical to Fas. Mouse Fas cDNA was also isolated by us from mouse macrophage cell line (Watanabe-Fukunaga et al. 1992b). The mouse Fas, consisting of 306 amino acids, has an identity of 49.3% with human Fas on the amino acid sequence level.

A comparison of the amino acid sequence of Fas with all protein sequences in GenBank indicated that Fas is a member of the TNF/NGF receptor family (Itoh et al. 1991; Nagata 1995). The members of this family include Fas, two TNF receptors (types I or 55K, and type II or 75K, respectively; Schall et al. 1990; Smith et al. 1990), the low-affinity NGF receptor (Johnson et al. 1986), the B-cell antigen CD40 (Stamenkovic et al. 1989), the T-cell antigen OX40 (Mallett et al. 1990), CD27 (Camerini et al. 1991), 4-1BB (Kwon and Weissman 1989), and the Hodgkin's lymphoma cell surface antigen CD30 (Dürkop et al. 1992; Fig. 1). The extracellular regions of the members in this family are rich in cysteine residues, and they can be divided into three to six subdomains. The amino acid sequence in this region is relatively conserved (about 24–30% identity), whereas the cytoplasmic region is not, except for some similarity between Fas and the TNF type I receptor (Itoh et al. 1991). The TNF and NGF receptors are receptors for cytokines, whereas Fas, CD40, CD27, and CD30 were identified as proteins which are recognized by specific monoclonal antibodies. Molecular cloning of the ligands for CD40, CD27, CD30, and 4-1BB (Armitage et al. 1992; Goodwin et al. 1993a, b; Smith et al. 1993) indicated that they are TNF-related type II-membrane proteins, and constitute a novel cytokine family (Farrah and Smith 1992). As described below, the Fas ligand is also a member of the TNF family.

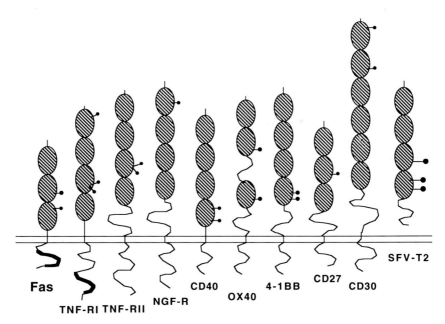

Fig. 1. The Fas/TNF/NGF receptor family. Members of the TNF/NGF receptor family are schematically shown. These include Fas, TNF type I and type II receptors, low-affinity NGF receptor, B cell antigen CD40, T cell antigens OX40, 4-1BB, and CD27, Hodgkin's lymphoma antigen CD30, and the soluble protein coded by Shope fibroma virus (SFV-T2). The *dashed regions* represent cysteine-rich subdomains, of which each member of the family contains 3–6. A domain of about 80 amino acids in the cytoplasmic regions of Fas and the type I TNF receptor has some similarity, and it is shown as a *bold line.* ♀ indicates N-glycosylation sites

2.2 Expression of Fas

The tissue distribution of the Fas mRNA has been examined in 8-week-old adult mice (Watanabe-Fukunaga et al. 1992b). The thymus, heart, liver, and ovary abundantly express the Fas mRNA, while little Fas mRNA was found in the brain, bone marrow, and spleen. In the thymus, most thymocytes except for double negative (CD4⁻ CD8⁻) thymocytes express Fas (Drappa et al. 1993; Ogasawara et al. 1993; J. Ogasawara, et al. 1995). Activated human T- and B-cells express Fas (Trauth et al. 1989), and lymphoblastoid cells transformed with human T-cell leukemia virus (HTLV-1; Debatin et al. 1990), human immuno-deficiency virus (HIV; Kobayashi et al. 1990) or Epstein-Barr virus (EBV; Falk et al. 1992) highly express Fas. Some other tumor cell lines such as human myeloid leukemia U937 (Yonehara et al. 1989), human squamous carcinoma CHU-2 (Itoh et al. 1991), and SV40-transformed mouse macrophage BAM3 cells (Watanabe-Fukunaga et al. 1992b) also express Fas, although the expression level is low compared with that of the lymphoblastoid cell lines.

The expression of Fas is up-regulated by interferon γ (IFN-γ) in the mouse macrophage BAM3, human adenocarcinoma HT-29, and mouse fibroblast L929 cell lines (Itoh et al. 1991; Watanabe-Fukunaga et al. 1992b), or by a combination of IFN-γ and TNFα in human tonsillar B cells (Möller et al. 1993).

3   Mutation in the Fas Gene of *lpr* Mice

3.1   Chromosomal Gene for Fas

Southern hybridization of genomic DNA indicated that there is only one chromo-somal gene for Fas in human and mouse chromosomes (Adachi et al. 1993). In situ hybridization localized the human gene on chromosome 10q24.1 (Inazawa et al. 1992), and interspecific backcross analysis indicated that the mouse Fas gene is in the region of chromosome 19, which is homologous to human 10q24.1 (Watanabe-Fukunaga et al. 1992b). Mouse Fas chromosomal gene was isolated, and its characterization indicated that mouse Fas gene is consisting of more than 70 kb, and split by 9 exons (R. Watanabe-Fukunaga and S. Nagata, unpubl. results).

3.2   Insertion of an Early Transposable Element in Intron of Fas Gene in lpr Mice

Referring the location of the mouse Fas gene to the mouse Genomic Database (GBASE), it was found that the Fas gene is close to the locus called *lpr* (lymphoproliferation; Watanabe et al. 1991). There are two known allelic muta-tions, *lpr* and *lpr^{cg}*, at the *lpr* locus. These mutants have a similar phenotype, but *lpr^{cg}* slightly complements the *gld* (generalized lymphoproliferative disease) mutation in double heterozygotes between *lpr* and *gld* mutations (Matsuzawa et al. 1990).

Northern hybridization of the thymus and liver from *lpr* mice showed little expression of the Fas mRNA (Watanabe-Fukunaga et al. 1992a). Accordingly, flow cytometry using anti-mouse Fas antibody hardly detected the Fas protein on the thymocytes from *lpr* mice (Drappa et al. 1993; Ogasawara et al. 1993). When the chromosomal DNA was analyzed by Southern hybridization, the Fas gene in *lpr* mice showed a distinct rearrangement. The Fas chromosomal gene was then molecularly cloned from *lpr* mice, and its structure was compared with that of the wild-type mice (Adachi et al. 1993). Restriction enzyme mapping of the Fas gene from *lpr* mice indicated that the promoter and exons of the Fas gene in this mouse are intact. However, an early transposable element (ETn) of 5.4 kb was inserted in intron 2 of the Fas gene. The ETn is a mouse endogenous retrovirus, of which about 1000 copies can be found in the mouse genome (Brulet et al. 1983). Although the ETn does not carry a meaningful open reading frame, it has long terminal repeat (LTR) sequences (about 300 bp) at both the 5' and 3' termini. This LTR sequence contains a poly(A) adenylation signal (AATAAA) which terminates the transcription at this region. In fact, short mRNAs of about 1 kb coding for exons 1 and 2 of the Fas gene are abundantly expressed in the thymus and liver of the *lpr* mice (Adachi et al. 1993). Furthermore, inserting the ETn into an intron of a mammalian expression vector dramatically, but not completely, reduced the expression efficiency in mammalian cells. These results indicate that in *lpr* mice, an insertion of an ETn into intron of the Fas gene greatly reduces the expression of the functional Fas mRNA, but its mutation is leaky. Later, several other groups reached the same conclusion by analyzing the Fas-transcript in *lpr* mice by means of the reverse polymerase-chain reaction (Chu et al. 1993; Kobayashi et al. 1993; Wu et al. 1993).

3.3 A Point Mutation in the Fas Gene of *lpr^{cg}* Mice

In contrast to the *lpr* mice, *lpr^{cg}* mice express the Fas mRNA of normal size as abundantly as the wild-type mice (Watanabe-Fukunaga et al. 1992a). However, characterization of the mRNA indicated that this mRNA carries a point mutation of T to A, which causes a replacement of isoleucine with asparagine in the Fas cytoplasmic region. This mutation is in the domain which has similarity with the TNF type I receptor (see below), and it abolishes the ability of Fas to transduce the apoptotic signal (Watanabe-Fukunaga et al. 1992a). Furthermore, when the corresponding amino acid (valine 238) of the human Fas was mutated to asparagine, it could not transduce the apoptotic signal into-cells (Itoh and Nagata 1993).

4 Fas-Mediated Apoptosis

4.1 Apoptosis In Vitro

As described above, the structure of the Fas indicated that Fas may be a receptor for a cytokine. However, it was not clear whether it is a receptor for a growth factor or a death factor. To assess the function of Fas, mouse cell transformants constitutively expressing human Fas were established using various mouse cell line as host cells (Itoh et al. 1991). When the transformed cells were treated with anti-human Fas antibody, cells expressing human Fas, but not the parental mouse cells, died within 5 h. Examination of the dying cells under an electron microscope revealed extensive condensation and fragmentation of the nuclei, which is characteristic of apoptosis. The chromosomal DNA started to degrade in a laddered fashion after a 2-h incubation with the anti-Fas antibody. A human Fas expression plasmid has also been introduced into a mouse IL-3 (interleukin-3)-dependent myeloid leukemia FDC-P1 cell line (Itoh et al. 1993). Although the transformed cells died due to IL-3 depletion, they did so over 36 h, as observed with the parental FDC-P1 cells. On the other hand, exposure to the anti-human Fas antibody killed the cells within 5 h in the presence of IL-3. From these results, we concluded that Fas actively mediates the apoptotic signal into cells, and the cytolytic anti-human Fas antibody works as agonist.

4.2 Apoptosis In Vivo

We established several hamster monoclonal antibodies against mouse Fas (Ogasawara et al. 1993). One of them had cytolytic activity in vitro. When this antibody is intraperitoneally injected into mice, the wild-type mice died within 5-6 h. On the other hand, neither lpr nor lpr^{cg} mice died by administration of anti-Fas antibody, indicating that the lethal effect of the anti-Fas antibody is due to binding of the antibody to the functional Fas in the tissues, and not due to a substance(s) such as endotoxin contaminated with the antibody. Furthermore, the fact that lpr^{cg} mice expressing the nonfunctional Fas are resistant to the lethal effect of the antibody suggests little involvement of the complement system in this killing process. Biochemical analysis of sera from the dying mice showed a specific and dramatic increase of GOT (glutamic oxaloacetic transaminase) and GPT (glutamic pyruvic transaminase) level shortly after injection of the antibody, suggesting the liver injury. Accordingly, histological analysis of the tissues indicated focal hemorrhage and necrosis in the liver (Fig. 2). On the other hand, when the dying hepatocytes were examined under the electron microscope, it showed a morphology characteristic of apoptosis (Fig. 2). These results indicate that individual hepatocytes died by apoptosis. However, since it occurred so rapidly and so widely, granulocytes and macrophages could not phagocytose the apoptotic cells, and the tissues went to the secondary necrosis.

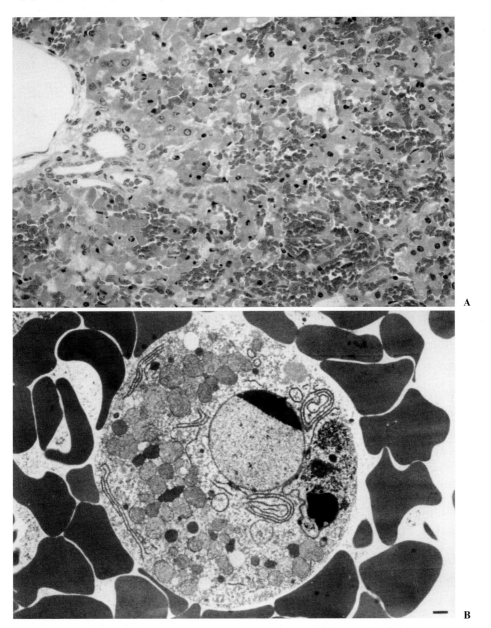

Fig. 2A, B. The Fas-mediated apoptosis of hepatocytes in vivo. The purified anti-mouse Fas antibody (100 μg) was subcutaneously injected into mice. At 2 h after injection, the liver section was stained with hematoxylin and eosin (**A**), which shows focal hemorrhage and necrosis. Only a few normal hepatocytes remained, and most hepatocytes carry pyknotic nuclei. **B** shows a liver section examined under a transmission electron microscope. The affected hepatocytes show the condensed and fragmented nuclei characteristic of apoptosis

The effect of the anti-Fas antibody in vivo seems to be a direct effect on the liver because the anti-Fas antibody also caused apoptosis in primary cultures of hepatocytes (Ni et al. 1994). These results indicate that the Fas expressed in mouse tissues (at least in the liver) is competent to transduce the apoptotic signal into cells.

4.3 Activation of Fas to Induce Apoptotic Signal

The apoptotic signal through Fas is induced by the binding of anti-Fas or anti-APO1 antibody, or the Fas ligand to Fas. The anti-human Fas antibody is an IgM class antibody which is an immunoglobulin pentamer, whereas the anti-APO1 antibody is an IgG_3 class antibody which tends to aggregate. The $F(ab')_2$ fragment or other isotypes of the anti-APO1 antibody hardly induces apoptosis of cells expressing Fas (Dhein et al. 1992). On the other hand, the cytotoxic activity of the inactive anti-APO1 antibody was reconstituted by cross-linking the antigen with a secondary antibody or with protein A. These results indicate that Fas dimerization alone is not sufficient to transduce the apoptotic signal. It seems that the oligomerization of at least three Fas molecules is a biologically relevant complex in generating an intracellular signal. As described below, the fact that Fas ligand is a TNF-related molecule which exists as a trimer (Smith and Baglioni 1987), agrees with this hypothesis.

4.4 Apoptotic Signal Mediated by Fas

Activation of Fas induces degradation of the chromosomal DNA within 3 h which eventually kills the cells within 5 h. These results suggest that a strong death signal is transduced from Fas. In some cells like mouse L929 and mouse primary hepatocytes, activation of Fas alone was not sufficient to induce the apoptotic signal. The presence of metabolic inhibitors such as cycloheximide or actinomycin D was required to induce the Fas-dependent apoptosis in these cells (Itoh et al. 1991). On the other hand, the activated or transformed T-cells can be killed by anti-Fas antibody alone. These results indicate that the signal-transducing machinery for Fas-induced apoptosis is present in most cells, and some cells express a labile protein(s) which works inhibitorily for Fas-mediated apoptosis. In fact, overexpression of the oncogene Bc12 product, which is known to inhibit apoptosis in various systems (Korsmeyer 1992), partially inhibited Fas-mediated apoptosis (Itoh et al. 1993).

The cytoplasmic domain of Fas consists of 145 amino acids, in which no motif for enzymatic activity such as kinases or phosphatase can be found (Itoh et al. 1991). However, about 70 amino acids in this region have significant similarity with a part of the cytoplasmic region of the type I, but not the type II TNF receptor (Itoh et al. 1991). TNF has numerous biological functions, including cytotoxic and proliferative activities (Old 1985). Tartaglia et al. (1991) have shown that the type I receptor is mainly responsible for the cytotoxic activity of TNF, while the type II receptor mainly mediates the proliferation signal in thymocytes. The similarity between Fas and the type I TNF receptor in their cytoplasmic regions therefore suggests an important role of this domain for apoptotic signal transduction. Analyses of serial deletions and point mutations in Fas have indicated that the domain conserved between Fas and the type I TNF receptor is essential for the function of Fas (Itoh and Nagata 1993). Extensive mutational analysis of the human type I TNF receptor have also indicated that the domain homologous to Fas is responsible and sufficient for TNF-induced cytolytic activity (Tartaglia et al. 1993) which agrees with our conclusion.

In addition to the signal-transducing domain, the mutational analysis of Fas revealed an inhibitory domain for apoptosis in the C-terminus. That is, a Fas mutant lacking 15 amino acids from the C-terminus was an up-mutant, in which about ten times less anti-Fas antibody than that required for the wild-type Fas was sufficient to induce apoptosis (Itoh and Nagata 1993). Moreover, even in L929 cells, activation of Fas alone (without metabolic inhibitors) was sufficient to induce apoptosis. It is possible that association of inhibitory molecule(s) mentioned above or modification of Fas at this region down-regulates the activity of Fas to transduce the apoptotic signal.

5 Fas Ligand, a Death Factor

5.1 Identification and Purification of Fas Ligand

Rouvier et al. (1993) have established a CTL hybridoma cell line (PC60-d10S, abbreviated to d10S) which has cytotoxic activity against thymocytes from the wild-type, but not *lpr* mice, suggesting the presence of a Fas ligand on its surface. To confirm the expression of Fas ligand in this cell line, we prepared a soluble form (Fas-Fc) of Fas by fusing the extracellular region of Fas to the Fc region of human IgG. The fusion protein inhibited the Fas-dependent CTL activity of d10S cells in a dose-dependent manner, and the Fas ligand was detected by FACS on the cell surface of d10S cells using labeled Fas-Fc (Suda and Nagata 1994). A subline of d10S which abundantly expresses the Fas ligand was established by repeated sorting procedure on FACS. After sorting 16 times, the subline (d10S16) expressed about 100 times moree Fas ligand than the original d10S cells and showed about 100 times more cytotoxic activity. The Fas ligand was then purified from d10S16 to homogeneity by affinity chromatographies using Fas-Fc and Con A. The purified Fas ligand had M_r of 40 kDa, and had specific cytolytic activity against cells expressing Fas (Suda and Nagata 1994), suggesting that a single protein (Fas ligand) is sufficient to induce apoptotic cell death by binding to Fas.

5.2 Molecular Properties of the Fas Ligand

A cDNA library was constructed from the twice sorted subline of d10S cells which abundantly express the Fas ligand, and the Fas ligand cDNA was isolated by the panning procedure using mFas-Fc as probe (Suda et al. 1993). The recombinant Fas ligand expressed in COS cells could kill the cells expressing Fas by apoptosis. The amino acid sequence deduced from the nucleotide sequence of the cDNA indicated that the Fas ligand is a type II membrane protein consisting of 278 amino acids (Suda et al. 1993).

Comparison of the amino acid sequence of the Fas ligand with all protein sequences in GenBank indicated that the Fas ligand is a member of the TNF family. As shown in Fig. 3, members of the TNF family include the Fas ligand, TNF, lymphotoxin (LT), and ligands for CD40, CD30, CD27, and 4-1BB. TNF was originally identified as a soluble cytokine (Pennica et al. 1984), which works as a trimer (Smith and Baglioni 1987). However, it was later shown that TNF is synthesized as a type II membrane protein which can be cleaved to produce a soluble form (Kriegler et al. 1988). LT consists of $LT\alpha$ and $LT\beta$, and is expressed in certain CTL (Androlewicz et al. 1992). $LT\alpha$, also called $TNF\beta$, is produced as a soluble cytokine with a signal sequence (Gray et al. 1984), while $LT\beta$ is a type II membrane protein (Browning et al. 1993). $LT\alpha$ and $LT\beta$ associate on the cell surface probably as a trimer, and bind to a receptor which has not yet been identified (Androlewicz et al. 1992). The ligands for CD40, CD30, CD27, and 4-1BB are type II membrane proteins expressed in activated T cells (Armitage et al. 1992; Goodwin et al. 1993a, b; Smith et al. 1993). When the Fas ligand was overproduced in COS cells, the soluble form of the Fas ligand which can actively induce apoptosis can be found in supernatant (Suda et al. 1993). These results

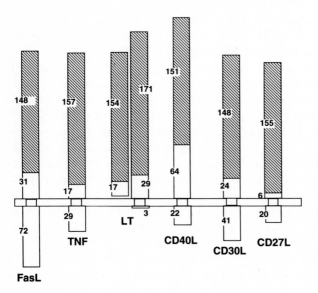

Fig. 3. The TNF family. The members of the TNF family are schematically shown. The members include the Fas ligand (*FasL*), membrane-bound *TNF*, Lymphotoxin (*LT*) which consists of LTα and LTβ, CD40 ligand (*CD40L*), CD30 ligand (*CD30L*) and CD27 ligand (*CD27L*). *Dashed regions* have significant similarity. *Numbers* indicate the amino acid number of the conserved, spacer, and intracellular regions

suggest that under abnormal conditions, the soluble form of the Fas ligand can be produced in the body as found in the TNF system (Old 1985).

The tertiary structure of TNF has been extensively studied. It forms an elongated, antiparallel β-pleated sheet sandwich with a jelly-roll topology (Eck and Sprang 1989; Eck et al. 1992; Banner et al. 1993). The significant conservation of the amino acid sequence among members suggests that others of the family, including Fas ligand, have a structure similar to TNF, and work as a trimer. However, despite the high similarity of the Fas ligand with TNF (about 30% identity of the amino acid sequence level), Fas ligand does not bind to the TNF receptor (Suda et al. 1993).

5.3 Expression of the Fas Ligand

Northern hybridization analysis of rat tissues with the cloned Fas ligand cDNA indicated that the Fas ligand is expressed abundantly in the testis, moderately in the small intestines, and weakly in the lung, whereas little expression of the Fas ligand mRNA was observed in the liver, heart, and ovary, where Fas is abundantly expressed.

In accord with the expression of the Fas ligand in the CTL cell line of d10S, activation of splenocytes with phorbol myristic acetate (PMA) and ionomycin strongly induced the expression of the Fas ligand mRNA (Suda et al. 1993), whereas the expression level of the Fas ligand mRNA in thymocytes was relatively weak even after activation with PMA and ionomycin.

6 Physiological Roles of the Fas System

6.1 Involvement of the Fas System in Development of T-Cells

As described above, the Fas gene is the structural gene for the *lpr* mutation. Since the mice homozygous at the *lpr* locus develop lymphadenopathy and suffer from autoimmune disease (Cohen and Eisenberg 1991), it is clear that Fas plays an important role in the development of T-cells. However, it remains controversial at which step of T-cell development Fas is involved. Immature T-cells are killed by apoptosis at least in two steps during development in the thymus (Ramsdell and Fowlkers 1990). Those T-cells carrying T-cell receptors which do not recognize self-MHC antigens as a restriction element are killed or "neglected", while the T-cells recognizing the self-antigens are killed by a process called negative selection. Analysis of thymic T-cell development in wild-type and *lpr* mice has suggested that the "neglected" thymocytes escape from apoptosis in the thymus of *lpr* mice, then migrate to the periphery (Zhou et al. 1993). On the other hand, Herron et al. (1993) reported that the development of T cell in the thymus is relatively normal in *lpr* mice. These different observations may be partly due to the leakiness of the *lpr* mutation as mentioned above. In addition to being expressed in thymocytes, Fas is expressed in activated mature T-cells (Trauth et al. 1989), and the prolonged activation of T-cells leads the cells susceptible against cytolytic activity of anti-Fas antibody (Owen-Schaub et al. 1992; Klas et al. 1993). Since mature T-cells from *lpr* mice are resistant against anti-CD3-stimulated suicide, Russell et al. (1993) have suggested a role of Fas-mediated apoptosis in the induction of peripheral tolerance and/or in the antigen-stimulated suicide of mature T-cells.

6.2   *Involvement of the Fas System in CTL-Mediated Cytotoxicity*

The Fas ligand is expressed in some CTL cell lines and in activated splenocytes (Suda et al. 1993), suggesting an important role of the Fas system in CTL-mediated cytotoxicity. Two mechanisms for CTL-mediated cytotoxicity are known (Golstein et al. 1991; Podack et al. 1991; Apasov et al. 1993). The one is a Ca^{2+}-dependent pathway in which perforin plays an important role. The other pathway is a Ca^{2+}-independent pathway, the mechanism of which is not well understood. In the perforin-null mice, the spleen cells still showed some Ca^{2+}-independent CTL activity, which was shown to be due to the Fas ligand (Kägi et al. 1994).

Mice carrying the *gld* mutation show phenotypes similar to *lpr* (Cohen and Eisenberg 1991). From bone-marrow transplantation experiments between *lpr* and *gld* mice, Allen et al. (1990) suggested that *gld* and *lpr* are mutations of an interacting pair of molecules. As shown above, the Fas gene is the structural gene for *lpr*, and Fas is the receptor for Fas ligand. It was now shown that *gld* mice carry a mutation in the Fas ligand gene (Takahashi et al. 1994). The mutation in Fas or Fas ligand causes lymphadenopathy and autoimmune disease, and the Fas ligand was found in CTL. These results imply that the Fas/Fas ligand system involved in the T-cell development plays an important role in CTL-mediated cytotoxicity. Moreover, it suggests that the killing process of autoreactive T-cells in T-cell development and the killing process of tumor cells by CTL may proceed by a similar mechanism.

6.3   *Pathological Tissue Damage Caused by the Fas System*

Fas is expressed not only in the thymocytes and lymphocytes, but also in other tissues such as the liver, heart, and lung (Watanabe-Fukunaga et al. 1992b). Although these organs are rather stable, and no apparent abnormal phenotypes are seen in these tissues of *lpr* mice, Fas may also be involved in development and/or turnover in these tissues. Since abnormal activation of Fas (administration of anti-Fas antibody) causes severe tissue damage (Ogasawara et al. 1993), as described above, it is possible that the Fas system is involved in many human autoimmune diseases such as fulminant hepatitis. In this regard, it is notable that a particular CTL cell line induces apoptosis in hepatocytes, which leads to fulminant hepatitis (Chisari 1992; Ando et al. 1993). As schematically shown in

Fig. 4. A model for the Fas-mediated cytotoxicity of CTL. A proposed mechanism for the Fas-mediated cytotoxicity in the CTL system is schematically shown. The target cell express the self, tumor, or virus antigen as a complex with MHC, which interacts with the T cell receptor (*TcR*) on CTL. This interaction activates the CTL, and induces the expression of the Fas ligand (*Fas-L*) gene. The Fas-L expressed on the cell surface of the CTL then binds to Fas on the target cells, and induces its apoptosis

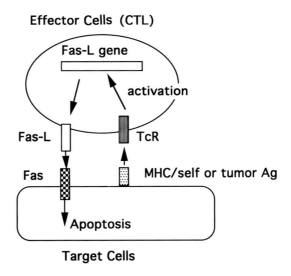

Fig. 5. Fas-mediated apoptosis. Fas and the Fas ligand are schematically shown. The Fas ligand binds to Fas on the cell surface probably as a trimer, and activates apoptotic signal transduction. In the cytoplasmic region of Fas, a region of about 80 amino acids is responsible for the signal transduction, while the C-terminal domain (about 15 amino acids) inhibits apoptosis

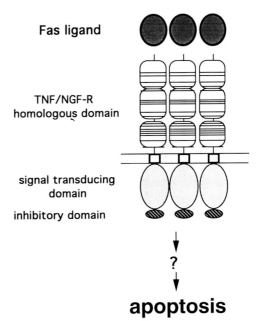

Fig. 4, the hepatocytes infected with HBV or HCV virus express the virus antigens as a complex with MHC. The interaction of CTL with the transformed hepatocytes may activate the CTL through the T-cell receptor, and induce the expression of Fas ligand gene. The Fas ligand then binds to Fas on the target cells, causing apoptosis. If involvement of the Fas system in human diseases is proven, antagonistic antibodies against Fas or Fas ligand, or the soluble form of Fas could be used in a clinical setting.

7 Perspectives

We demonstrated that Fas ligand is a death factor, and Fas is its receptor. These results indicate that just as growth factor and its receptor regulate cell proliferation, apoptosis is also regulated by a death factor and its receptor (Fig. 5). It would be interesting to examine what kinds of signals are transduced through Fas to induce apoptosis. The gain-of-function mutation of the growth factor system causes cellular transformation, whereas the loss-of-function mutation of the Fas system (*lpr* mutation) causes lymphadenopathy. In this regard, Fas and the Fas ligand may be considered as tumor suppressor genes. The loss-of-function mutation in the growth factor system causes the disappearance or dysfunction of specific cells. As pointed out above, abnormal activation (gain-of-function mutation) of the Fas or Fas ligand may cause fulminant hepatitis or other diseases such as CTL-mediated autoimmune diseases.

Acknowledgments. I thank Drs. O. Hayaishi and C. Weissmann for encouragement and discussion. The work was carried out in a collaboration with Drs. T. Suda, J. Ogasawara, T. Takahashi, M. Adachi, N. Itoh and R. Watanabe-Fukunaga, and supported in part by Grants-in-Aid from the Ministry of Education, Science and Culture of Japan. I also thank Ms. K. Mimura for secretarial assistance.

References

Adachi M, Watanabe-Fukunaga R, Nagata S (1993) Aberrant transcription caused by the insertion of an early transposable element in an intron of the Fas antigen gene of *lpr* mice. Proc Natl Acad Sci USA 90: 1756–1760

Allen RD, Marshall JD, Roths JB, Sidman CL (1990) Differences defined by bone marrow transplantation suggest that *lpr* and *gld* are mutations of genes encoding an interacting pair of molecules. J Exp Med 172: 1367–1375

Ando K, Moriyama T, Guidotti LG, Wirth S, Schreiber RD, Schlicht HJ, Huang S, Chisari FV (1993) Mechanisms of class I restricted immunopathology. A transgenic mouse model of fulminant hepatitis. J Exp Med 178: 1541–1554

Androlewicz MJ, Browning JL, Ware CF (1992) Lymphotoxin is expressed as a heteromeric complex with a distinct 33-kDa glycoprotein on the surface of an activated human T cell hybridoma. J Biol Chem 267: 2542–2547

Apasov S, Redegeld F, Sitkovsky M (1993) Cell-mediated cytotoxicity: contact and secreted factors. Curr Opin Immunol 5: 404–410

Arai K-I, Lee F, Miyajima A, Miyatake S, Arai N, Yokota Y (1990) Cytokines: coordinators of immune and inflammatory responses. Annu Rev Biochem 59: 783–836

Armitage RJ, Fanslow WC, Strockbine L, Sato TA, Clifford KN, Macduff BM, Anderson DM, Gimpel SD, Davis-Smith T, Maliszewski CR, Clark EA, Smith CA, Grabstein KH, Cosman D, Spriggs MK (1992) Molecular and biological characterization of a murine ligand for CD40. Nature 357: 80–82

Banner DW, D'Arcy A, Janes W, Gentz R, Schoenfeld H-J, Broger C, Loetscher H, Lesslauer W (1993) Crystal structure of the soluble human 55 kd TNF receptor-human TNFβ complex: implication for TNF receptor activation. Cell 73: 431–445

Browning JL, Ngam-ek A, Lawton P, DeMarinis J, Tizard R, Chow EP, Hession C, O'Brine-Greco B, Foley SF, Ware CF (1993) Lymphotoxin β, a novel member of the TNF family that forms a heteromeric complex with lymphotoxin on the cell surface. Cell 72: 847–856

Brulet P, Kaghad M, Xu Y-S, Croissant O, Jacob F (1983) Early differential tissue expression of transposon-like repetitive DNA sequences of the mouse. Proc Natl Acad Sci USA 80: 5641–5645

Camerini D, Walz G, Loenen WAM, Borst J, Seed B (1991) The T cell activation antigen CD27 is a member of the nerve growth factor/tumor necrosis factor receptor gene family. J Immunol 147: 3165–3169

Chisari FV (1992) Hepatitis B virus biology and pathogenesis. Mol Gen Med 2: 67–103

Chu BJ-L, Drappa J, Parnassa A, Elkon KB (1993) The defect in *Fas* mRNA expression in MRL/*lpr* mice is associated with insertion of the retrotransposon, *ETn*. J Exp Med 178: 723–730

Cohen PL, Eisenberg RA (1991) *Lpr* and *gld*: single gene models of systemic autoimmunity and lymphoproliferative disease. Annu Rev Immunol 9: 243–269

Debatin K-M, Goldmann CK, Bamford R, Waldmann TA, Krammer PH (1990) Monoclonal-antibody-mediated apoptosis in adult T-cell leukaemia. Lancet 335: 497–500

Dhein J, Daniel PT, Trauth BC, Oehm A, Möller P, Krammer PH (1992) Induction of apoptosis by monoclonal antibody anti-APO-1 class switch variants is dependent on cross-linking of APO-1 cell surface antigens. J Immunol 149: 3166–3173

Drappa J, Brot N, Elkon KB (1993) The Fas protein is expressed at high levels on CD4$^+$CD8$^+$ thymocytes and activated mature lymphocytes in normal mice but not in the lupus-prone strain, MRL *lpr/lpr*. Proc Natl Acad Sci USA 90: 10340–10344

Dürkop H, Latza U, Hummel M, Eitelbach F, Seed B, Stein H (1992) Molecular cloning and expression of a new member of the nerve growth factor receptor family that is characteristic for Hodgkin's disease. Cell 68: 421–427

Eck MJ, Sprang SR (1989) The structure of tumor necrosis factor-α at 2.6 Å resolution. J Biol Chem 264: 17595–17605

Eck MJ, Ultsch M, Rinderknecht E, de Vos AM, Sprang SR (1992) The structure of human lymphotoxin (tumor necrosis factor-β) at 1.9-Å resolution. J Biol Chem 267: 2119–2122

Ellis RE, Yuan J, Horvitz HR (1991) Mechanisms and functions of cell death. Annu Rev Cell Biol 7: 663–698

Falk MH, Trauth BC, Debatin K-M, Klas C, Gregory CD, Rickinson AB, Calender A, Lenoir GM, Ellwart JW, Krammer PH, Bornkamm GW (1992) Expression of the APO-1 antigen in Burkitt lymphoma cell lines correlates with a shift towards a lymphoblastoid phenotype. Blood 79: 3300–3306

Farrah T, Smith CA (1992) Emerging cytokine family. Nature 358: 26

Golstein P, Ojcius DM, Young JD-E (1991) Cell death mechanisms and the immune system. Immunol Rev 121: 29–65

Goodwin RG, Alderson MR, Smith CA, Armitage RJ, VandenBos T, Jerzy R, Tough TW, Schoenborn MA, Davis-Smith T, Hennen K, Falk B, Cosman D, Baker E, Sutherland GR, Grabstein KH, Farrah T, Giri JG, Beckmann MP (1993a) Molecular and biological characterization of a ligand for CD27 defines a new family of cytokines with homology to tumor necrosis factor. Cell 73: 447–456

Goodwin RG, Din WS, Davis-Smith T, Anderson DM, Gimpel SD, Sato TA, Maliszewski CR, Brannan CI, Copeland NG, Jenkins NA, Farrah T, Armitage RJ, Fanslow WC, Smith CA (1993b) Molecular cloning of a ligand for the inducible T cell gene 4-1BB: a member of an emerging family of cytokines with homology to tumor necrosis factor α. Eur J Immunol 23: 2631–2641

Gray PW, Aggarwal BB, Benton CV, Bringman TS, Henzel WJ, Jarrett JA, Leung DW, Moffat B, Ng P, Svedersky LP, Palladino MA, Nedwin GE (1984) Cloning and expression of cDNA for human lymphotoxin, a lymphokine with tumour necrosis activity. Nature 312: 721–724

Herron LR, Eisenberg RA, Roper E, Kakkanaiah VN, Cohen PL, Kotzin BL (1993) Selection of the T cell receptor repertoire in *Lpr* mice. J Immunol 151: 3450–3459

Inazawa J, Itoh N, Abe T, Nagata S (1992) Assignment of the human Fas antigen gene (FAS) to 10q24.1. Genomics 14: 821–822

Itoh N, Nagata S (1993) A novel protein domain required for apoptosis: mutational analysis of human Fas antigen. J Biol Chem 268: 10932–10937

Itoh N, Yonehara S, Ishii A, Yonehara M, Mizushima S, Sameshima M, Hase A, Seto Y, Nagato S (1991) The polypeptide encoded by the cDNA for human cell surface antigen Fas can mediate apoptosis. Cell 66: 233–243

Itoh N, Tsujimoto Y, Nagata S (1993) Effect of bcl-2 on Fas antigen-mediated cell death. J Immunol 151: 621–627

Johnson D, Lanahan A, Buck CR, Sehgal A, Morgan C, Mercer E, Bothwell M, Chao M (1986) Expression and structure of the human NGF receptor. Cell 47: 545–554

Kägi D, Vignaux H, Ledermann B, Bürki K, Depraetere V, Nagata S, Hengartner H, Golstein P (1994) Fas and perforin pathway as major mechanism of Tall-mediated cytotoxicity. Science 265: 528–530

Klas C, Debatin K-M, Jonker RR, Krammer PH (1993) Activation interferes with the APO-1 pathway in mature human T cells. Int Immunol 5: 625–630

Kobayashi N, Hamamoto Y, Yamamoto N, Ishii A, Yonehara M, Yonehara S (1990) Anti-Fas monoclonal antibody is cytocidal to human immunodeficiency virus-infected cells without augmenting viral replication. Proc Natl Acad Sci USA 87: 9620–9624

Kobayashi S, Hirano T, Kakinuma M, Uede T (1993) Transcriptional repression and differential splicing of Fas mRNA by early transposon (*ETn*) insertion in autoimmune *LPR* mice. Biochem Biophys Res Commun 191: 617–624

Korsmeyer SJ (1992) Bcl-2 initiates new category of oncogenes: regulators of cell death. Blood 80: 879–886

Kriegler M, Perez C, DeFay K, Albert I, Lu SD (1988) A novel form of TNF/Cachectin is a cell surface cytotoxic transmembrane protein: Ramifications for the complex physiology of TNF. Cell 53: 45–53

Kwon BS, Weissman SM (1989) cDNA sequences of two inducible T-cell genes. Proc Natl Acad Sci USA 86: 1963–1967

Mallett S, Fossum S, Barclay AN (1990) Characterization of the MRC OX40 antigen of activated CD4 positive T lymphocytes – a molecule related to nerve growth factor receptor. EMBO J 9: 1063–1068

Matsuzawa A, Moriyama T, Kaneko T, Tanaka M, Kimura M, Ikeda H, Katagiri T (1990) A new allele of the *lpr* locus, *lprcg*, that complements the *gld* gene in induction of lymphadenophaty in the mouse. J Exp Med 171: 519–531

Möller P, Henner C, Leithäuser F, Eichelmann A, Schmidt A, Brüderlein S, Dhein J, Krammer PH (1993) Co-regulation of the APO-1 antigen with ICAM-1 (CD54) in tonsillar B cells and coordinate expression in follicular center B cells and in follicle center and mediastinal B cell lymphomas. Blood 81: 2067–2075

Nagata S (1993) Apoptosis-mediating Fas antigen and its natural mutation. In: Tomci LD, Cope FC (eds) Cold Spring Harbor Lab Press, Cold Spring Harbor, New York PP. 313–326

Ni R, Tomita Y, Matsuda K, Ichihara A, Ishimura K, Ogasawara J, Nagata S (1994) Fas-mediated apoptosis in primary cultured mouse hepatocytes. Exp Cell Res 215: 332–337

Oehm A, Behrmann I, Falk W, Pawlita M, Maier G, Klas C, Li-Weber M, Richards S, Dhein J, Trauth BC, Ponstingl H, Krammer PH (1992) Purification and molecular cloning of the APO-1 cell surface antigen, a member of the tumor necrosis factor/nerve growth factor receptor superfamily: sequence identity with the Fas antigen. J Biol Chem 267: 10709–10715

Ogasawara J, Watanabe-Fukunaga R, Adachi M, Matsuzawa A, Kasugai T, Kitamura Y, Itoh N, Suda T, Nagata S (1993) Lethal effect of the anti-Fas antibody in mice. Nature 364: 806–809

Ogasawara J, Suda T, Nagata S (1995) Selective apoptosis of CD4 CD8 thymocytes by the anti-Fas antibody. J Exp Med 181: 485–491

Old LJ (1985) Tumor necrosis factor (TNF). Science 230: 630–632

Owen-Schaub LB, Yonehara S, Crump III WL, Grimm EA (1992) DNA fragmentation and cell death is selectively triggered in activated human lymphocytes by Fas antigen engagement. Cell Immunol 140: 197–205

Pennica D, Nedwin GE, Hayflick JS, Seeburg PH, Derynck R, Palladino MA, Kohr WJ, Aggarwal BB, Goeddel DV (1984) Human tumour necrosis factor: precursor structure, expression and homology to lymphotoxin. Nature 312: 724–728

Podack KR, Hengartner H, Lichtenheld MG (1991) A central role of perforin in cytolysis? Annu Rev Immunol 9: 129–157

Raff MC (1992) Social controls on cell survival and cell death. Nature 356: 397–400

Ramsdell F, Fowlkers BJ (1990) Clonal deletion versus clonal anergy: the role of the thymus in inducing self tolerance. Science 248: 1342–1348

Rouvier E, Luciani M-F, Golstein P (1993) Fas involvement in Ca^{2+}-independent T cell-mediated cytotoxicity. J Exp Med 177: 195–200

Russell JH, Rush B, Weaver C, Wang R (1993) Mature T cells of autoimmune *lpr/lpr* mice have a defect in antigen-stimulated suicide. Proc Natl Acad Sci USA 90: 4409–4413

Schall TJ, Lewis M, Koller KJ, Lee A, Rice GC, Wong GHW, Gatanaga T, Granger GA, Lentz R, Raab H, Kohr WJ, Goeddel DV (1990) Molecular cloning and expression of a receptor for human tumor necrosis factor. Cell 61: 361–370

Smith CA, Davis T, Anderson D, Solam L, Beckman MP, Jerzy R, Dower SK, Cosman D, Goodwin RG (1990) A receptor for tumor necrosis factor defines an unusual family of cellular and viral proteins. Science 248: 1019–1023

Smith CA, Gruss H-J, Davis T, Anderson D, Farrah T, Baker E, Sutherland GR, Brannan CI, Copeland NG, Jenkins NA, Grabstein KH, Gliniak B, McAlister IB, Fanslow W, Alderson M, Falk B, Gimpel S, Gillis S, Din WS, Goodwin RG, Armitage RJ (1993) CD30 antigen, a marker for Hodgkin's lymphoma, is a receptor whose ligand defines an emerging family of cytokines with homology to TNF. Cell 73: 1349–1360

Smith RA, Baglioni C (1987) The active form of tumor necrosis factor is a trimer. J Biol Chem 262: 6951–6954

Stamenkovic I, Clarck EA, Seed B (1989) A B-lymphocyte activation molecule related to the nerve growth factor receptor and induced by cytokines in carcinomas. EMBO J 8: 1403–1410

Suda S, Nagata S (1994) Purification and characterization of the Fas-ligand that induces apoptosis. J Exp Med 179: 873–878

Suda S, Takahashi T, Golstein P, Nagata S (1993) Molecular cloning and expression of the Fas-ligand, a novel member of the tumor necrosis factor family. Cell 75: 1169–1178

Takahashi T, Tanaka M, Brannan CI, Jenkins NA, Copeland NG, Suda T, Nagata S (1994) Generalized lymphoproliferative disease in mice, caused by a point mutation in the Fas ligand. Cell 76: 969–976

Tartaglia LA, Weber RF, Figari IS, Reynolds C, Palladino MA Jr, Goeddel DV (1991) The two different receptors for tumor necrosis factor mediate distinct cellular responses. Proc Natl Acad Sci USA 88: 9292–9296

Tartaglia LA, Ayres TM, Wong GHW, Goeddel DV (1993) A novel domain within the 55 kd TNF receptor signals cell death. Cell 74: 845–853

Trauth BC, Klas C, Peters AMJ, Matzuku S, Möller P, Falk W, Debatin K-M, Krammer PH (1989) Monoclonal antibody-mediated tumor regression by induction of apoptosis. Science 245: 301–305

Walker NI, Harmon BV, Gobe GC, Kerr JFR (1988) Patterns of cell death. Methods Achiev Exp Pathol 13: 18–54

Watanabe T, Sakai Y, Miyawaki S, Shimizu A, Koiwai O, Ohno K (1991) A molecular genetic linkage map of mouse chromosome-19, including the *lpr*, *Ly-44*, and *TdT* genes. Biochem Genet 29: 325–336

Watanabe-Fukunaga R, Brannan CI, Copeland NG, Jenkins NA, Nagata S (1992a) Lympho-proliferation disorder in mice explained by defects in Fas antigen that mediates apoptosis. Nature 356: 314–317

Watanabe-Fukunaga R, Brannan CI, Itoh N, Yonehara S, Copeland NG, Jenkins NA, Nagata S (1992b) The cDNA structure, expression, and chromosomal assignment of the mouse Fas antigen. J Immunol 148: 1274–1279

Wu J, Zhou T, He J, Mountz JD (1993) Autoimmune disease in mice due to integration of an endogenous retrovirus in an apoptosis gene. J Exp Med 178: 461–468

Wyllie AH, Kerr JFR, Currie AR (1980) Cell death: the significance of apoptosis. Int Rev Cytol 68: 251–306

Yonehara S, Ishii A, Yonehara M (1989) A cell-killing monoclonal antibody (anti-Fas) to a cell surface antigen co-downregulated with the receptor of tumor necrosis factor. J Exp Med 169: 1747–1756

Zhou T, Bluethmann H, Eldridge J, Berry K, Mountz JD (1993) Origin of CD4⁻CD8⁻B220⁺ T cells in MRL-*lpr/lpr* mice. J Immunol 150: 3651–3667

Myc-Mediated Apoptosis

Y. Kuchino, A. Asai, and C. Kitanaka

Abstract

Mammalian cells contain an intron-less *myc* gene, such as the rat s-*myc* gene and human *myc* L2 gene, which are expressed in rat embryo chondrocytes and human testis, respectively. Our recent findings demonstrated that s-Myc expression suppresses the growth activity and tumorigenicity of glioma cells, indicating that s-Myc acts as a negative regulator in tumor growth. In addition, we found that s-Myc overexpression can effectively induce apoptotic cell death in human and rat glioma cells without serum deprivation, which is distinct from c-Myc-mediated apoptosis.

1 Introduction

The *myc* gene was initially isolated as a transforming gene of an avian myelo-cytomatosis virus (Ivanov et al. 1964; Sheiness et al. 1978). Since then, seven cellular genes with significant homology to the viral *myc* gene (v-*myc*) have been found in mammalian cells (c-*myc*, N-*myc*, L-*myc*, B-*myc*, s-*myc*, *myc*L2, N-*myc*2). The *myc* family genes except B-*myc* encode nuclear phosphoproteins that contain a basic helix-loop-helix leucine zipper (bHLH-Zip) motif required for DNA binding and protein dimerization (Blackwood and Eisenman 1991; Prendergast et al. 1991). To form a sequence-specific DNA-binding complex, Myc family proteins associate with a heterologous bHLH protein Max (Blackwood et al. 1992). Myc-Max heterodimers specifically bind the CACGTG sequence and function as transcriptional activation factors (Kretzner et al. 1992). In addition, the c-*myc* gene is responsible in the early stage of the cell cycle depending upon stimuli from mitogens. Expression of the c-Myc protein generally stimulates cell cycle progression toward the S phase. Thus, for the past several years, we have thought that products of the *myc* family genes such as c-Myc probably act as growth-promoting factors (Luscher and Eisenman 1990; Penn et al. 1990).

High levels of *myc* gene expression are frequently observed in a variety of tumors such as retinoblastomas, neuroblastomas, Burkitt lymphomas, and small cell lung carcinomas (Schwab 1988). Furthermore, c-Myc has been shown to

Biophysics Division, National Cancer Research Institute, Tsukiji 5-1-1, Chuo-ku, Tokyo 104, Japan cooperate with mutated Ras to transform rat embryo fibroblasts. These findings suggest that Myc may also play a pivotal role in transformation of various cells.

In contrast to these positive roles of Myc in cell proliferation and tumor formation, Wurm et al. (1986) found that increased levels of c-Myc are cytotoxic to mammalian cells. Pallavicini et al. (1990) investigated the relationship between expression levels of c-Myc and their cellular consequences, and showed that high levels of c-Myc are inversely correlated with cell growth. A similar opposite role of Myc in cell proliferation was also found by Evan et al. (1992). Their finding indicated that enhanced c-*myc* expression can induce apoptotic cell death of rat and mouse fibroblast cells whose cell proliferation was arrested by deprivation of serum growth factors.

Apoptosis occurs frequently in various tissues and cells on removal of specific growth factors and cytokines, or on addition of physiological regulatory hormones (Arends and Wyllie 1991; Ellis et al. 1991; Raff 1992; Wyllie 1992; Schwartz et al. 1993). Many of these factors presumably send signals from the cell surface to the nucleus, resulting in transcription of apoptotic genes and causing initial events in the induction of apoptosis. Since these multiple signals may be able to regulate the expressions of various genes such as *myc* and *fos*, there may be multiple ways to initiate the apoptotic pathway (Williams 1991; Oren 1992; Owens and Cohen 1992; Marx 1993; Vaux 1993). On the contrary, in the final stage of the apoptotic pathway, several common alterations such as progressive fragmentation of chromosomal DNAs and nuclear condensation are induced in cells. Based on these considerations, Even et al. (1992) proposed that although the commitments of individual cells to apoptosis are probably determined by a variety of stochastic factors, c-Myc may be directly involved in initiation of apoptosis to organize the common processes of apoptosis.

We summarize here the reports indicating that Myc plays a key role in induction of apoptotic cell death, and discuss the cell type specificity and molecular mechanism of Myc-mediated apoptosis.

2 c-Myc-Mediated Apoptosis

Evan et al. (1992) proposed that c-Myc can be a potent inducer of apoptosis in mammalian cells because they found that increased Myc expression due to gene transfection induced apoptosis in immortalized fibroblast cells such as Rat-1a and NIH3T3 cells when these cells were subjected to serum deprivation. Since this initial report on Myc-mediated apoptosis, there have been an increasing number of papers indicating that c-Myc has a key role in the apoptotic pathway. Recently, two groups successively reported that elevated levels of c-Myc protein in Chinese hamster ovary (CHO) cells are cytotoxic and induce apoptotic cell death. Bissonnette et al. (1992) demonstrated that the cell death induced in CHO cells by overexpression of c-Myc is due to apoptosis. To prove the c-Myc-dependent induction of apoptosis, they used a c-*myc* gene transcription system under the control of a heat shock promoter and demonstrated that c-Myc overexpression induced by incubation at 43°C resulted in cell death. Pallavicini et al. (1990) introduced also the c-*myc* gene connected with the *Drosophila* heat shock promoter into CHO cells and demonstrated that the presence of high levels of c-Myc was associated with a decreased rate of DNA synthesis and inversely correlated with cell survival postheating. Evan's group used a different induction system of c-Myc: they fused the c-*myc* gene to part of the human β-estrogen receptor (ER) gene and transfected the chimeric gene into Rat-1 cells (Evan et al. 1992). Using a Rat-1 transfectant (Rat-1/c-Myc-ER) in which the chimeric protein was activated only in the presence of β-estradiol, they showed that serum-deprived Rat-1/c-Myc-ER cells became arrested in a G_0 like state, and that addition of β-estradiol prevented the arrest of cellular progression and induced rapid apoptotic cell death. Their results suggest that cell cycle arrest induced in cells on withdrawal of serum growth factors may be involved in apoptosis induction.

Cell cycle arrest induced by serum starvation can be produced by other interventions. For instance, vitamin K3 is known to induce persistent c-*myc* expression in nasopharyngeal carcinoma cells, resulting in cell cycle arrest and apoptotic cell death of the cells. In this system, Wu et al. (1993) suggested that increased expression of c-Myc may play an important role in the signaling mechanism of VK3-induced cell death. Yonish-Rouach et al. (1991, 1993), and Lotem and Sachs (1993) also reported that upregulation of c-*myc* expression in cell cycle-blocked cells induces apoptosis. They introduced wild-type *p53* (wt-*p53*) gene into murine myeloid leukemic M1 cells having constitutive overexpression of c-*myc*, and found that wt *p53* expression resulted in induction of rapid cell death in M1 cells. wt p53 is known to cause G1 arrest in some cells. Therefore, it has been thought that this cell cycle arrest induced by wt p53 expression might cooperate with endogenous c-*myc* expression to initiate apoptosis. However, a recent finding that G1 arrest of mouse embryo fibroblasts by isoleucine starvation does not confer susceptibility to apoptosis suggests that cell cycle arrest is not sufficient to induce c-Myc-mediated apoptosis (Wagner et al. 1994). c-Myc-mediated apoptosis requires wt-p53 expression in a manner independent of cell cycle arrest. In fact, as reported by Hermeking and Eick (1994), c-Myc expression in mouse fibroblasts stabilizes the wt p53 protein causing accumulation of cellular content of the protein.

Several recent reports demonstrated that apoptosis is induced in myeloid cells in response to withdrawal of IL-3, which dramatically decreases c-*myc* expression in the cells and arrests the cell cycle at the G1 phase, but, interestingly, when constitutive c-Myc expression was induced in IL-3-deprived myeloid cells, the susceptibility of the cells to apoptotic cell death was remarkably enhanced. These findings strongly suggest that c-Myc also has a key role in the induction of apoptosis in myeloid cells (Askew et al. 1991; Cleveland et al. 1994; Malde and Collins 1994; Marvel et al. 1994).

Lymphocytes are activated by binding of antigens with receptors on the cell surface, or by exposure of the receptors to agents that induce the same signal as antigens. There are reports that immature T cells and some T cell hybridomas undergo apoptosis on activation through the CD3-T cell receptor complex.

Shi et al. (1992) demonstrated a role of c-*myc* expression in activation-induced T-cell apoptosis using c-*myc* antisense oligonucleotides. Their results indicated that antisense c-myc oligodeoxy-nucleotides inhibited the constitutive expression of c-Myc in T-cell hybridomas and interfered with all aspects of activation-induced T-cell apoptosis. In addition, they found that antisense c-myc oligodeoxy-nucleotides did not inhibit the induction of apoptosis in the same T-cell line by dexamethasone, and suggested from these results that there may be multiple apoptotic pathways.

Similar relationships between c-Myc overexpression and apoptosis have been observed in human Colo 320 cells and human promyelocytic leukemia HL-60 cells treated with teniposide (VM-26), which is a cancer chemotherapeutic drug with a high target specificity for DNA topoisomerase II (Bertrand et al. 1991; Tepper and Studzinski 1992). Bertrand et al. demonstrated secondary DNA fragmentation resembling chromatin endonucleolytic cleavage by apoptosis in HL-60 and Colo 320 cells, in both of which the c-*myc* gene is amplified, but not in human adenocarcinoma HT-29 cells, in which it is not amplified. From these findings, they suggested that c-*myc* overexpression may be involved in apoptotic DNA fragmentation.

Hoang et al. (1994) demonstrated that in c-Myc-overexpressing Rat-1a-myc fibroblasts that is undergoing apoptosis, the cyclin A mRNA level is elevated but not cyclin B, C, D1, and E transcripts, and that inducible cyclin A expression is sufficient to cause apoptosis. Based on these findings, they concluded that apoptosis induced in Rat-1a fibroblasts by c-Myc overexpression is associated with an elevated cyclin A mRNA level. On the other hand, Packham and Cleveland (1994) proposed the possibility that ornithine decarboxylase (ODC) is a mediator of c-Myc-induced apoptosis. Their observation indicated that apoptosis induced in IL-3-deprived murine myeloid cells by enforced c-*myc* expression is associated with the constitutive, growth factor-independent expression of ODC. In addition, they demonstrated that enforced expression of ODC following IL-3 withdrawal is sufficient to induce apoptosis in the cells.

In contrast, Davidoff and Mendelow (1993) concluded that upregulation of c-*myc* gene expression is not essential for activation of the apoptosis cascade because they observed induction of apoptosis in HL-60 cells exposed to Mafosfamide, which is a cyclophosphamide derivative, and found that the apoptosis was associated with marked decrease in c-*myc* gene transcription. Similar apoptosis accompanied by c-*myc* downregulation has been demonstrated in several other systems. Interleukin 2 (IL-2) and IL-3 are hematopoietic growth factors that enhance the survival of hematopoietic precursor cells and are important in regulation of hematopoiesis. Deprivation of IL-2 and IL-3 from T- or B-lymphocytes has been shown to cause c-*myc* repression and result in induction of apoptosis in these cells (Duke and Cohen 1986; Vaux et al. 1988; Vaux and Weissman 1993).

In contrast to IL-2 and IL-3, IL-6 induces growth inhibition and apoptosis in cells such as murine hematopoietic Y6 cells. IL-6 induces differentiation of Y6 cells into macrophages. Oritani et al. (1992) found that IL-6-induced differentiation of Y6 cells into macrophages followed by apoptosis is preceded by downregulation of the c-*myc* gene. Thus they speculated that repression of c-*myc* gene expression by addition of IL-6 might have a regulatory role in induction of apoptosis in hematopoietic cells. Alnemri et al. (1992) found that glucocorticoid also represses c-*myc* gene expression in human 697 pre-B lymphocytes causing apoptosis. By using antisense c-myc oligomers, Thulasi et al. (1993) demonstrated directly that c-*myc* repression is essential for glucocorticoid-induced human leukemic cell lysis.

3   s-Myc-Mediated Apoptosis

3.1   Structural Feature of the s-myc Gene

Most cellular *myc* family genes contain three exons separated by two introns and have a single long translational open reading frame. The open reading frames of these genes generally begin at the ATG codon near the 5′ terminal region of exon 2, extend through exon 3, and terminate in exon 3. In contrast, the s-*myc* gene isolated from a rat genomic library by screening with the exon 3 region of the v-*myc* gene as a probe is a unique member of the *myc* gene family in having a single open reading frame without introns (Sugiyama et al. 1989).

The s-*myc* gene contains chimeric structural features of mixed c-*myc* and N-*myc* genes (Fig. 1). For instance, mammalian c-*myc* gene contains two promoters used for regulation of transcriptional expression of the gene in a tissue-specific manner. The TATA box in both promoter regions of mammalian c-*myc* gene is followed about 30 bp downstream by an AC dinucleotide sequence which is a transcription initiation site. The s-*myc* gene contains at least six TATA sequences within the region 1 kbp upstream of a putative translation initiation site. Among them, the third TATA sequence is followed by an AC dinucleotide sequence located about 30 bp downstream. Interestingly, this AC sequence is followed by a sequence that is highly homologous with the sequence of rat N-*myc* exons 1 and 2. Contrary to the c-*myc* and s-*myc* genes, mammalian N-*myc* gene has no typical TATA sequence in the 5' upstream region of the transcription initiation sites.

In the exon 1 region, the c-*myc* gene has the signal sequence recognized by a *cis*-acting translation inhibition factor (Parkin et al. 1988). Comparison of the

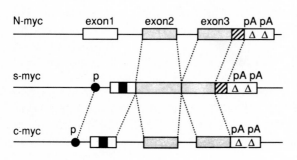

Fig. 1. Diagram of the structures of the *myc* family genes. Exons are indicated by boxes. Stippled boxes represent the protein coding regions. The positions of promoters and polyadenylation signals are shown by *solid circles* and *open triangles*, respectively. *Closed boxes* indicate the positions of the binding site of a *cis*-acting translation inhibition factor. *Hatched boxes* contain the conserved sequence required for integration of hepatitis B virus

sequence in the 5' noncoding region between the c-*myc* and s-*myc* genes showed that the s-*myc* gene contains a similar signal sequence preceded by the TATA sequence to that of the c-*myc* gene. Indeed, in vitro transcription-translation analyses using different plasmids containing the predicted open reading frame of the s-*myc* gene with 5' and 3' untranslated regions of various lengths demonstrated that deletion of a part of the 5' untranslated region from the s-*myc* transcripts increased the translation efficiency (Sugiyama et al. 1989). As shown in Fig. 2, a capped s-*myc* transcript synthesized from plasmid pTDD containing the shortest 5' untranslated region and the full length of the predicted open reading frame of the s-*myc* gene produced a significant amount of methionine-labeled protein of 55 kDa, whereas the translations of s-*myc* transcrips from pTEK, pTED, and pTETD, which contained almost the full length of the 5' untranslated region, resulted in greatly decreased formation of p55 protein. These results indicate that the translational efficiency of s-*myc* transcripts is markedly stimulated by deletion of the *Eco*T221-*Dra*I region, which has a sequence similar to that of the binding site of a *cis*-acting translation inhibition factor located in the c-*myc* exon 1 region (Fig. 3), from the 5' untranslated sequence.

The polyadenylation signal, AATAAA, required for transcription termination of the *myc* family genes, is present in the third exon. Mammalian c-*myc* gene

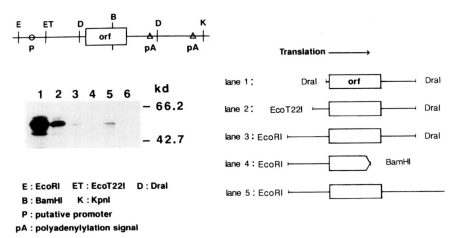

Fig. 2. In vitro translation of s-myc RNAs. The DNA fragments, 4.9 kbp *Eco*RI-*Kpn*I, 2.9 kbp *Eco*RI-*Dra*I, 2.3 kbp *Eco*T221-*Dra*I, and 2 kbp *Dra*I-*Dra*I were subcloned into the appropriate site of phagemid vector pTZ18R containing the bacteriophage T7 promoter to obtain plasmids pTEK, pTED, pTETD, and pTDD, respectively. Each plasmid DNA was linearized by *Hind* III digestion and then transcribed by T7 RNA polymerase in the presence of 7-methyl GTP. Each capped s-myc RNA synthesized in vitro was translated in vitro using a rabbit reticulocyte lysate with S^{35}-methionine. Finally, labeled polypeptides were analysed by SDS polyacrylamide gel electrophoresis. *Lanes 1–5* Translation products of RNAs from pTDD, pTETD, pTED, pTEK, and pTEK, respectively. For *lane 4*, the plasmid DNA, pTEK, was digested with *Bam*HI instead of *Hind* III to prepare the template RNA having a part of the protein coding region of the *s-myc* gene. *Lane 6* Translation products without template RNAs

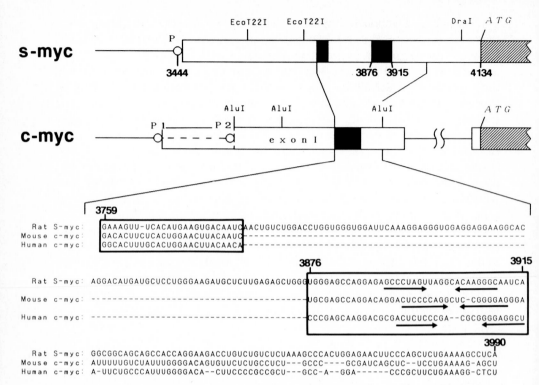

Fig. 3. Similarities of the nucleotide sequence for binding of a *cis*-acting translation inhibition factor between the c-*myc* and s-*myc* genes. The conserved nucleotide sequences in the region known to be required for translation inhibition are *boxed in the lower figure* (*black boxes in upper figures*). In the *upper figures, open boxes* and *hatched boxes* represent the 5'-untranslated region and the protein coding region, respectively. *P(0)* indicates the promoter. *ATG* indicated *above the hatched regions* shows the translation initiation site

contains dual polyadenylation signals in exon 3. In the mouse and human N-*myc* genes, there is a single polyadenylation signal, although the rat N-*myc* gene contains two polyadenylation signals. The s-*myc* gene contains two independent polyadenylation signals in the 3' noncoding region as well as the c-*myc* gene.

Another characteristic structure of the s-*myc* gene is a unique sequence which may be required for integration of hepatitis virus in the 3' untranslated region. Initially, this sequence was found in the exon 3 region of the woodchuck N-*myc* gene (Fourel et al. 1990).

3.2 Structural Feature and Biological Functions of the s-Myc Protein

The protein encoded by the s-*myc* gene consists of 429 amino acids and shows 60 and 31% sequence homology with the rat N- and c-Myc proteins at the amino acid level, respectively. However, the s-Myc protein contains all the consensus amino acid sequences conserved in the c- and N-Myc proteins except the sequence context required for phosphorylation by casein kinase II (CKII) in the internal acidic region (domain IV; Fig. 4; Meek and Street 1992).

In the N-terminal portion, the s-Myc protein contains two phosphorylation sites, Thr-58 and Ser-62, required for high level of transactivation of gene expression as well as c-Myc and N-Myc. Transient transcription assay using an artificial GAL4-s-Myc fusion protein demonstrated that the N-terminal segment of the s-Myc protein fused to the GAL4 DNA-binding domain was more effective than the c-Myc protein in stimulating transcription of the reporter gene (Fig. 5; Asai et al. 1994b). In the C-terminal portion, the s-Myc protein contains the bHLH-Zip structure, which is involved in DNA binding and dimerization with Max and highly conserved in the Myc family proteins. Together with the recent observations, indicating that the full length s-Myc protein can activate transcription of the gene containing the CACGTG sequence in the promoter region (Kitanaka et al. 1995), these structural features clearly indicate that the s-Myc protein is a nuclear DNA-binding protein and acts as a transactivation factor.

The Myc protein is generally phosphorylated at three distinct regions (Meek and Street 1992). CK II specifically phosphorylates serine and threonine residues located in highly acidic regions of proteins (domains IV, VI). The presence of glutamate or aspartate residues adjacent to the carboxyterminal of the amino acid residues that are phosphorylation sites is also especially important as a substrate for CK II. This substrate specificity of CK II suggests that the serine residues

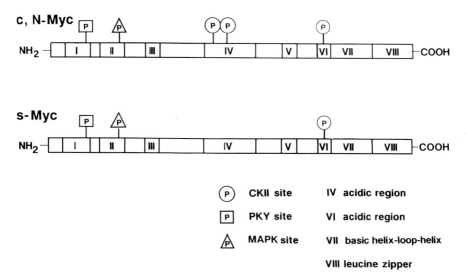

Fig. 4. Alignment of homologous amino acid sequences of mouse and rat Myc family proteins. The amino acid sequences deduced from the DNA sequences reported so far are aligned to achieve maximum homology. *Boxes* show the conserved amino acid sequences in the proteins

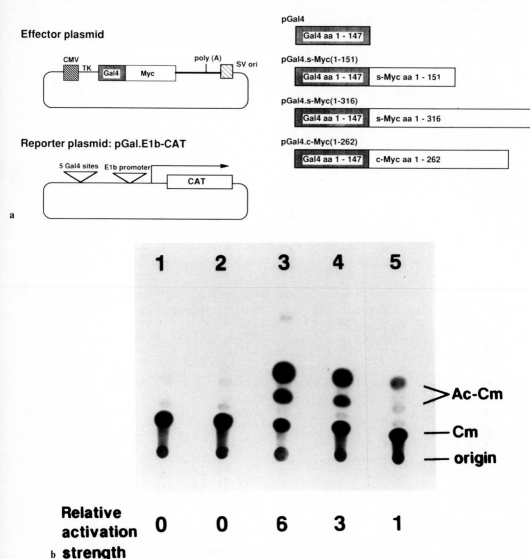

located in an internal acidic region (domain IV) and in a segment flanking the basic DNA-binding domain (domain VI) can be phosphorylated by CK II. The s-Myc protein also contains two serine residues at the same positions in the domain IV region as those in the c-Myc protein (Fig. 6). However, those serine residues are not followed by acidic amino acid-rich sequence. Therefore, these serine residues may not be phosphorylated by CK II. In contrast, domain VI of the s-Myc protein contains a serine residue followed by a DLED sequence, in which the serine residue could be phosphorylated by CK II.

As there are several reports that phosphorylation of the serine residues in domain IV by CKII is important for cell transformation (Stone 1987), it is supposed that s-Myc may have no transforming activity. Indeed, constitutive

Domain IV

```
                                    * *
rat c-Myc          HEETPPTTSSDSEEEQDDEEEIDVV
rat N-Myc          LSTSGEDTLSDSDDEDDEEEDEEEEIDVV
mouse L-Myc        QACSGSESPSDSEGEEIDVV
woodchuck N-Myc2   LSTSGEDALSDEVDEEEDEEEEIDVV

rat s-Myc          LSSSLEDFLSNSGYVEEGGEEI
```

Fig. 6. Amino acid sequence conserved in the domain IV region of Myc family proteins. *Asterisks in the boxed region* (domain IV) show phosphorylation sites of CK II

expression of s-Myc selectively suppresses the growth activity of neural tumor cells and alters the transformed phenotypes of the tumor cells including their tumorigenicity in nude mice (Sugiyama et al. 1989; Kuchino et al. 1990; Asai et al. 1994a). Moreover, we have recently demonstrated that s-Myc expression inhibits cell cycle progression of neural tumor cells at the G1/S boundary, although the s-Myc protein has a potential to act as a transcription factor as well as the c-Myc protein (Asai et al. 1994b; Kitanaka et al. 1995). These biological functions of the s-Myc protein are contrary to the positive functions of the c-Myc protein in cell proliferation, but resemble the functions of the wt-p53 protein, whose expression arrests the cell cycle at the G1 phase resulting in growth suppression.

◄——

Fig. 5a, b. Transactivation activity of s-Myc fused to GAL4 DNA binding domain. **a** Diagrams of activator and reporter plasmids used for cotransfection in analysis of the CAT activity. *H: Hind* III; *X: Xba* I; *B: Bam*HI; *Ac-Cm*, acetylated forms of chloramphenicol; *Cm* chloramphenicol; *CMV* cytomegalovirus promoter; *Tk* herpes simplex virus thymidine kinase leader sequence; *GAL4* a coding sequence of GAL4 DNA binding domain (amino acids 1 to 147); *poly (A)* rabbit β-globin polyadenylation site; *SV ori* sequence of the replication origin of SV40; *E1b* promoter sequence of adenovirus *E1b* gene. **b** CAT enzyme activity analyzed by thin-layer chromatography. The activator plasmid contains the cytomegalovirus promoter linked to the 5'-untranslated leader sequence of herpes simplex virus thymidine kinase (HSV *tk*) gene and a gene fragment encoding the GAL4 DNA binding domain (amino acids 1–147) followed by the rabbit β-globin polyadenylation signal. A gene fragment encoding the amino terminal region of s-Myc (amino acids 1–151 or 1–316) or of human c-Myc (amino acids 1–262) was inserted between the *gal4* gene and the polyadenylation signal. The reporter plasmid contains five copies of the GAL4-binding sequences upstream of the adenovirus *E1b* promoter with a single TATA sequence which regulates transcription initiation of the *cat* gene followed by the simian virus 40 polyadenylation signal. Reporter plasmid DNA (5 µg) was cotransfected with activator plasmid DNA (5 µg) into human glioma U251 cells growing exponentially in Dulbecco's modified Eagle medium containing 10% fetal calf serum by the lipofectin-mediated transfection method. The efficiency of transfection was standardized by cotransfection of pSVβ-gal plasmid DNA (10 µg), and using β-galactosidase Enzyme Assay System (Promega), the β-galactosidase activity was measured. The cells were harvested 48 h after transfection of the plasmid DNAs and then the CAT activity in the cells was determined as described by Gorman et al. (1982). Radioactivity of each spot was estimated by BAS 2000 Image Analyzer (Fujix, Tokyo). The relative transactivation activity was represented by ratio of each CAT activity to that of pGAL4c-Myc. *Lane 1* pMAMneo + reporter; *lane 2* pGAL4 + reporter; *lane 3* pGAL4-smyc(1–151) + reporter; *lane 4* pGAL 4-smyc (1–316) + reporter; *lane 5* pGAL4-cmyc (1–262) + reporter

3.3   Expression of the s-myc Gene

Most eukaryotic genes without introns have been regarded as transcriptionally inactive pseudogenes. However, surprisingly, Northern blot analysis probed by [32]p-labeled s-*myc* gene fragment demonstrated a weak expression of 3.2 kb s-myc mRNA in the head and neck of rat embryo (Fig. 7a, lane 7). Since expression of s-myc mRNA in rat embryo brain including cerebrum and cerebellum was not observed (data not shown), the synthesis of s-myc mRNA in rat embryo tissues such as the cartilage of the skull base and sternum was expected. As shown in Fig. 8b (lanes 1–3), definite expression of 3.2 kb s-myc m RNA was detected in the

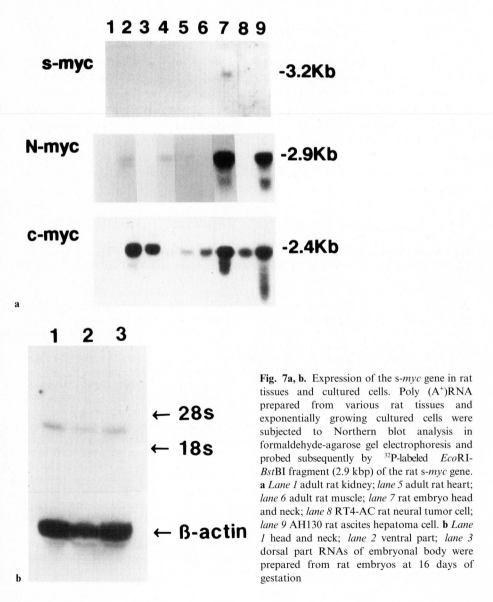

Fig. 7a, b. Expression of the s-*myc* gene in rat tissues and cultured cells. Poly (A[+])RNA prepared from various rat tissues and exponentially growing cultured cells were subjected to Northern blot analysis in formaldehyde-agarose gel electrophoresis and probed subsequently by ^{32}P-labeled *Eco*RI-*Bst*BI fragment (2.9 kbp) of the rat s-*myc* gene. **a** *Lane 1* adult rat kidney; *lane 5* adult rat heart; *lane 6* adult rat muscle; *lane 7* rat embryo head and neck; *lane 8* RT4-AC rat neural tumor cell; *lane 9* AH130 rat ascites hepatoma cell. **b** *Lane 1* head and neck; *lane 2* ventral part; *lane 3* dorsal part RNAs of embryonal body were prepared from rat embryos at 16 days of gestation

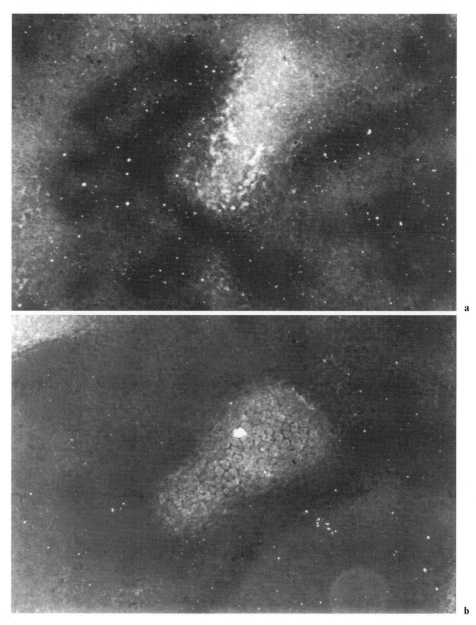

Fig. 8a, b. Expression of s-myc mRNA in rat embryo cartilage bones. **a, b** In situ hybridization analysis of s-*myc* gene expression with antisense and sense probes. The antisense and sense s-*myc* oligodeoxynucleotides consisting of 41 nts (nucleotides 4924 to 4964 of the protein coding region of the s-*myc* gene) were chemically synthesized and labeled at their 3′ terminals in the presence of a-^{35}S-dATP (1200 Ci/mmol, Dupont/NEN) using terminal deoxynucleotidyl transferase (Dupont). The in situ hybridizations according to the protocol recommended by the manufacturer (Dupont) were performed on 12-μm-thick cryostat sections of the vertebral body of a frozen rat embryo at 16 days of gestation. Finally, the sections were immersed in photoemulsion and served for autoradiography and observed under dark field microscopy cartilages not only of the head and neck, but also of ventral and dorsal parts of the body including the vertebral sternal bones. In situ hybridization using an anti-sense riboprobe was carried out to confirm s-myc mRNA expression in embryo cartilage (Fig. 8). The synthesis of s-myc mRNA in rat embryo cartilage was further confirmed by RNase protection analysis. A 208 nucleotide (nts) band was detected with antisense s-myc riboprobe in rat embryo head and neck, spine and ribs, but not in the distal extremities of rat embryo (Fig. 9). Also clear staining was observed in the same peripheral regions of the vertebral bones and the sternal bones when the same consecutive sections were stained immunohisto-chemically with s-Myc polyclonal antibody (Fig. 10). Fractionation study of cellular compo-nents indicated that most of the synthesized s-Myc protein are localized in the nuclei of cells (Fig. 11). These observations showed that the s-*myc* gene is an active gene being transcribed in rat embryo cartilage and the protein encoded by the s-*myc* gene is a nuclear protein (Asai et al. 1994b).

This is the first demonstration that one of the Myc family proteins is selectively expressed in chondrocytes located in the peripheral region of rat embryo carti-lage, although c-Myc expression playing a positive role in cell proliferation has been observed in chondrocytes of the proximal growth plates of rat bones (Farquharson et al. 1992). Normal expression of the s-*myc* gene has not been observed in various cell lines and adult rat tissues examined so far (Sugiyama et al. 1989). Therefore, this finding provided a possibility that embryo chondro-cytes may contain a unique transcription factor required for the selective expres-sion of the s-*myc* gene.

Recent reports showed that hypertrophic chondrocytes are the cells undergo-ing programmed cell death, in which selective expression of protein-glutamine γ-glutamyltransferase ("tissue" transglutaminase) has been observed (Aeschlimann et al. 1993; P.J.A. Davies, pers. comm.). The "tissue" transglutaminase which

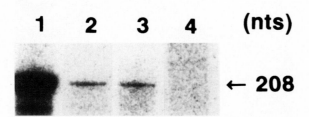

Fig. 9. RNase protection analysis of s-myc mRNA. Total poly(A⁺)RNA prepared from rat embryo tissues subjected to RNase protection analysis. A DNA fragment corresponding to nucleotides 4763 to 4970 of the s-Myc protein coding region, which contains the sequence encoding the s-Myc specific unique peptide with the sequence context for phosphorylation by CKII, was subcloned into a plasmid vector pTZ18R. Using a commercial transcription kit (Promega) with T7 RNA polymerase (Boehringer Mannheim) and α-³²P-UTP (800 Ci/mmol, Dupont/NEN), ³²P-labeled antisense s-myc riboprobe of 281 nts was synthesized. Total poly(A⁺)RNA (30 μg) was hybridized for 18 h to 8 × 10⁴ dpm of antisense riboprobe and then digested with RNase A and T1 as recommended by the manufacturer (Promega). Separation by SDS polyacrylamide gel electrophoresis containing 8% urea was then performed. *Lane 1* s-*myc* transfectant; *lane 2* rat embryonal head and neck; *lane 3* spine and ribs; *lane 4* embryonal extremities

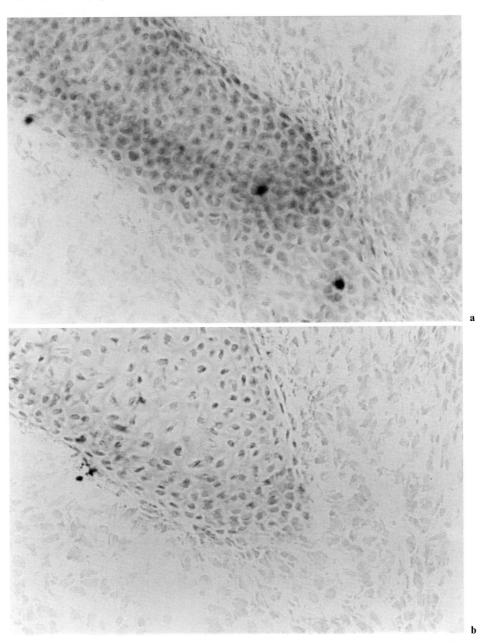

Fig. 10a, b

catalyzes the synthesis of isopeptide cross-linked proteins with γ-glutaminyl-lysyl isopeptide bonds is induced and activated during many forms of apoptosis, and is accumulated in apoptotic cells (Fesus et al. 1991). The possibility that s-Myc functions as a transcription factor recognizing the specific DNA sequence as well

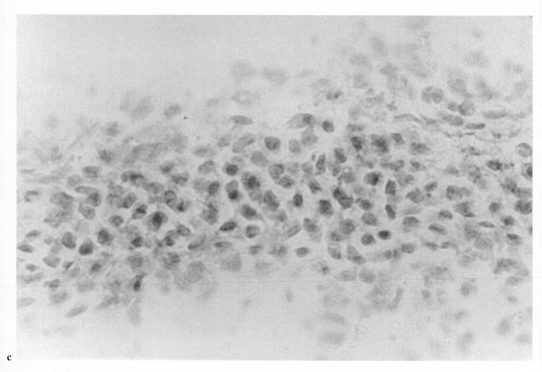

c

Fig. 10a–c. Immunohistochemical analysis of s-Myc in cartilage bones of a rat embryo at 16 days of gestation. Fresh frozen rat embryos at 16 days of gestation were sectioned to 8 μm by cryostat. Prior to immunostaining, samples were fixed in 4% paraformaldehyde and incubated in methanol containing 0.3% hydroxyperoxide for 30 min to block the endogenous peroxidase activity. After washing with PBS, the sections were blocked with 5% normal goat serum (Vector Lab) and then incubated with either rabbit anti-s-Myc polyclonal antibody against the synthetic peptide consisting of 15 amino acids positions 224 to 238 of the s-Myc protein (corresponding to nucleotides 4803–4848) or with nonimmune rabbit serum for 1 h. They were washed and incubated for 1 h with biotinylated goat anti-rabbit IgG (Vector Lab.). The sections were washed and incubated with peroxidase-conjugated avidin-biotin complex (Vector Lab.) and treated with 0.1% diaminobenzidine with 0.02% H_2O_2 for 1–5 min followed by counterstaining with hematoxylin. **a** The vertebral body with s-Myc polyclonal antibody: **b** The section containing the same microscopic field as **a** with nonimmune rabbit serum. **c** The sternal bone with s-Myc antibody

as the c-Myc protein was proved by a series of experiments using the full-length s-Myc protein (Kitanaka et al. 1995). Therefore, the fact that the s-Myc protein was expressed in the cells being committed to differentiate to hypertrophic chondrocytes which undergo programmed cell death may provide positive evidence indicating that the s-Myc protein may play a role in the in vivo regulation of the transcription expression of the gene(s) such as the "tissue" transglutaminase gene involved in apoptosis.

On the other hand, the studies using transgenic mice have demonstrated that the protein encoded by the proto-oncogene c-*fos* is not required for the growth of most cell types but is involved in development, including programmed cell death

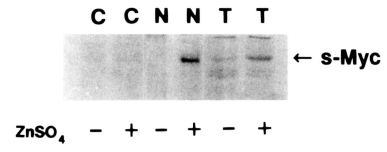

Fig. 11. Localization of the s-Myc protein synthesized in U251smyc10 cell. Eight hours after addition of zinc sulfate, U251smyc10 cells were harvested and fractionated to nuclear and cytoplasmic components by 0.4% NP-40 and centrifugation (6000 g). Nuclear and cytoplasmic lysates were separately prepared. Fifteen micrograms of each lysate were loaded in each lane and separated by electrophoresis using a 10% polyacrylamide gel with 0.1% SDS. Proteins were electrophoretically transferred to a nitrocellulose membrane. This membrane was subjected to Western blot analysis probed by s-Myc polyclonal antibody. After washing with PBS, the filter was incubated with biotinylated goat anti-rabbit IgG. The filter was then further incubated in avidin-biotin complex (ABC) solution containing alkaline phosphatase-conjugated streptoavidin-biotin complex (Vector Lab.) and subsequently in substrate solution (Bio-Rad Lab.). Letters *C*, *N*, and *T* designate cytoplasmic, nuclear, and total lysates, respectively. Characters – and + represent pre- and post-addition of zinc sulfate, respectively

of several distinct tissues such as the cartilage and neural tissues (Johnson et al. 1992; Smeyne et al. 1993). This finding raised another possibility that the s-Myc protein may function in transcription regulation of the apoptotic gene(s) in chondrocytes in cooperation with other gene products such as the c-Fos protein.

3.4 s-Myc-Mediated Apoptosis

The finding that the s-Myc protein was expressed in the cells being committed to differentiate to hypertrophic chondrocytes which undergo programmed cell death strongly suggests that the s-Myc protein could have a crucial role in apoptosis induction. In addition, recent reports showed that overexpression of either the c-Myc protein or the wt p53 protein can induce apoptosis in several types of cells including immortalized fibroblast cells and colon cancer cells (see above; Shaw et al. 1992; Clarke et al. 1993; Lowe et al. 1993). Moreover, induction of apoptosis was recently reported to be frequently associated with growth arrest of cells at the G1 phase (see Sect. 1). Prompted by these findings, the ability of the s-Myc protein as a member of the Myc family to induce apoptosis was examined. For this purpose, the s-*myc* gene linked with the human metallothionein promoter was transfected into rat glioma cells by the lipofectin-mediated transfection method. After selection of G418-resistant clones, s-*myc* gene expression was induced in transfectants by addition of zinc sulfate. Northern blot analysis with an s-*myc* gene fragment as a probe revealed that s-*myc* expression in s-*myc* transfectant U251smyc10 from human glioma U251 cells without zinc

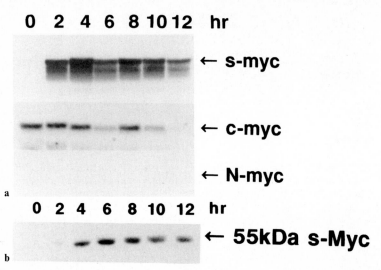

Fig. 12a, b. Expression of the s-*myc* gene products in s-*myc* transfectants by addition of zinc sulfate.
a Northern blot analysis of s-myc mRNA with the 2.9-kbp fragment of the s-*myc* gene as a probe.
Poly(A⁺)RNA (2 μg) from U251smyc10 cells collected at 2-h intervals after addition of 200 μM zinc
sulfate was hybridized with a nick-translated ^{32}P-labeled s-*myc* probe. The same filter was subsequently hybridized with ^{32}P-labeled *Bg*1III-*Hpa*I fragment (823 bp) of the rat c-*myc* exon 3 region
and *Hinc*II-*Pst*I fragment (590 bp) of the rat N-myc exon 3 region as probes. **b** Western blot analysis
of the s-Myc protein synthesized in s-*myc* transfectants with s-Myc polyclonal antibody. Crude
lysates were prepared from U251smyc10 cells as described by Blackwood et al. (1992) at 2-h intervals
after induction by 200 μM zinc sulfate. Ten Micrograms of the crude lysate were loaded in each lane.
Western blot analysis was carried out as described in Fig. 11

sulfate treatment was not detectable, and yet a visible band was observed when
induced by addition of 200 μM zinc sulfate. The maximum level of s-*myc* expression was observed at 4 h after induction, and gradually decreased with time
(Fig. 12a). The level of c-myc and β-actin mRNAs remained unchanged by
induction of s-Myc synthesis and addition of zinc sulfate (data not shown).
Synthesis of the s-Myc protein in U251smyc10 cells was also verified by Western
blot analysis (Fig. 12b). The analysis with anti-s-Myc antibody demonstrated
that accumulation of the s-Myc protein was presented at 4 h post-induction, and
remained high for at least 20 h in the zinc sulfate-treated transfectants. Similar
induction of the s-*myc* gene products was also observed in s-*myc* transfectants,
9Lsmyc7 from rat 9L glioma cells and C6smyc4 from rat C6 glioma cells,
although 160 μM zinc sulfate was used for s-*myc* induction in these rat glioma
cells (data not shown).

Fig. 13a, b. Electron microscopic analysis of U251smyc10 cells undergoing apoptotic cell death by
induction of s-Myc expression. **a** Transmission electron micrographs of control U251ter1 cells which
are transfected with pSVneoHMTer DNA without the s-*myc* gene, 24 h after addition of 200 μM zinc
sulfate. **b** U251smyc10 cells, 24 h after addition of the same amount of zinc sulfate

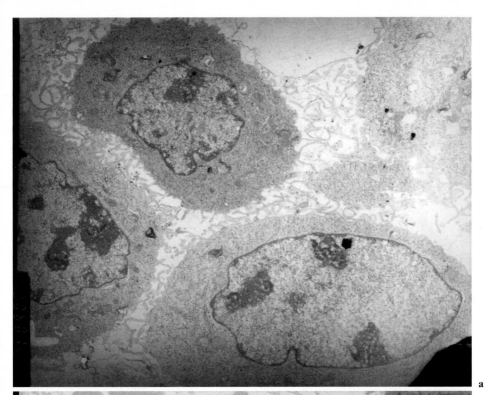

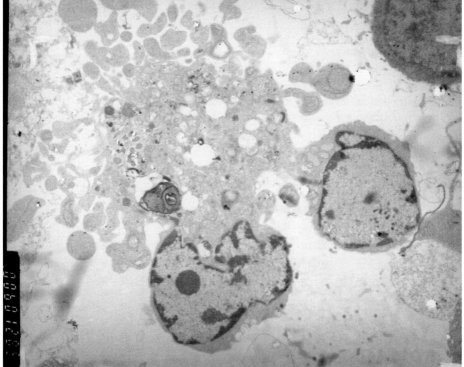

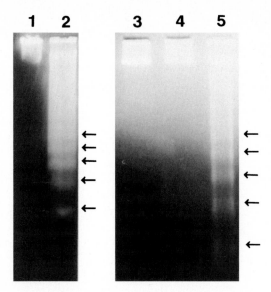

Fig. 14. Gel electrophoresis showing genomic DNA fragmentation in s-*myc* transfectants treated with zinc sulfate. *Lanes 1–5* DNA fragmentation patterns of U251ter1, U251smyc10, 9L, 9Lter3, and 9Lsmyc7 cells. Cells were harvested 48 h after induction by zinc sulfate and lysed in 100 mM Tris-HCl buffer (pH 7.5) containing 400 µg/ml proteinase K, 150 mM NaCl, and 20 mM EDTA for 3 h at 37 °C. The lysate was digested with RNase A. Ten micrograms of genomic DNA were electrophoresed on a 2% agarose gel and stained with ethidium bromide. Each *arrow* corresponds to 180, 360, 540, 720, and 900 bp from *bottom to top*

Microscopic analysis indicated that glioma cells started to detach from the plastic dishes within 6 h after s-Myc induction. Three days after s-Myc induction, almost all U251smyc10 cells died and became floaters. To determine whether cell death is due to apoptosis or necrosis, the morphological changes and the fragmentation of chromosomal DNA in s-Myc-induced glioma cells were analyzed. Typical morphological changes in apoptosis, such as condensation of the nucleus, chromatin localization to the periphery of the nucleus, and formation of translucent cytoplasmic vacuoles (Arends and Wyllie 1991), were detected under an electron microscope (Fig. 13a, b). Progressive degradation of chromosomal DNA with a ladder of DNA fragments was observed after s-Myc induction (Fig. 14), which is also typical of apoptosis (Arends and Wyllie 1991). Similar findings of apoptosis were observed in C6myc4 cells. However, it has been known that zinc ion acts as an inhibitor of endonucleases causing the chromosomal DNA degradation associated with apoptosis. In order to exclude the effect of zinc sulfate on apoptosis, we carried out colony formation assays using pCEP expression vector. This assay provided clear results suggesting that expression of the s-*myc* gene subcloned into the vector induced apoptotic cell death in human and rat glioma cells without zinc sulfate (Kitanaka et al. 1995).

These results demonstrated that the s-Myc protein acts as an inducer of apoptosis even in the presence of serum growth factors. As described above, s-Myc is expressed in rat embryo chondrocytes which are committed to differentiate into hypertrophic chondrocytes that undergo programmed cell death. Taken together, these results strongly suggest that the s-Myc protein may be involved in the induction of apoptosis not only in in vitro systems, but also under physiological conditions (Asai et al. 1994b). If this is so, s-Myc protein expression in vivo may have a crucial role in the elimination of hypertrophic chondrocytes during the remodeling of rat embryonal bones.

4   Escape from Myc-Induced Apoptotic Cell Death

The *bcl-2* gene is a proto-oncogene isolated from the breakpoint of transactivation between human chromosomes 14 and 18 found in human follicular B-cell lymphomas (Tsujimoto et al. 1984; Strasser et al. 1990). There are several recent reports that *Bcl-2* itself has no transforming activity, but prolongs cell survival and prevents cell death (Tsujimoto, this Vol.).

Deprivation of IL-2 or IL-3, which causes downregulation of c-*myc* gene expression in T- or B-lymphocytes, may be directly involved in suppression of cellular proliferation and induction of apoptosis (Duke and Cohen 1986; Askew et al. 1991; Cleveland et al. 1994; Malde and Collins 1994; Marvel et al. 1994). For instance, when IL-3-dependent pre-B lymphocytes are deprived of IL-3, they cease to proliferate and enter a cytostatic phase before finally dying through apoptosis. Bcl-2 expression blocks this process and results in promotion of hemopoietic cell survival (Hockenbery et al. 1990; Vaux and Weissman 1993). Cleveland et al. (1994) reported that v-Raf also has the ability to suppress cell death of myeloid 32D.3 cells following the withdrawal of IL-3. In this study, they suggested that v-Raf kinase promotes Bcl-2 function and that v-*raf* clones decreased the requirements for IL-3 for growth. A similar process of apoptosis preceded by downregulation of the c-*myc* gene is observed in glucocorticoid-treated human pre-B leukemia cells. Alnemri et al. (1992) suggested that glucocorticoid-induced apoptosis of lymphocytes may be very similar to IL-3-deprived cell death of pre-B cells and that, in both systems, c-*myc* downregulation may be directly involved in the inhibition of proliferation and cell death of lymphocytes of B origin. Moreover, they showed that Bcl-2 blocked glucocorticoid-induced apoptosis of huamn pre-B leukemias, prolonging their survival when the level of c-*myc* expression is repressed. Similar suppression by Bcl-2 of the induction of programmed cell death of leukemic cells by deregulated c-*myc* expression was observed by Lotem and Sachs (1993). Interestingly, they found that a tumor-promoting phorbol ester and mutant p53 expression can also suppress this programmed death. More direct evidence that Bcl-2 expression blocks apoptotic cell death induced by c-Myc expression has been provided by others. Bissonnette et al. (1992) demonstrated that cell death induced in CHO cells by overexpression of the c-*myc* gene under the control of a heat-shock promoter was suppressed by constitutive expression of the human *bcl-2* gene connected with the promoter of spleen focus-forming virus. Similar suppression of c-Myc-induced apoptosis in CHO cells was observed by expression of Mcl-1, which has sequence similarity to Bcl-2 (Reynolds et al. 1994).

Fanidi et al. (1992) transfected the c-*myc* gene linked to the human estrogen receptor gene into Rat-1 cells and showed that apoptosis induced in Rat-1 cells by addition of β-estradiol was prevented by retrovirus-directed constitutive expression of the human *bcl-2* gene. Wagner et al. (1993) reported that ectopic expression of Bcl-2 specifically blocks apoptosis induced in serum-deprived Rat-1a fibroblasts by constitutive c-*myc* expression. Bcl-2 expression also inhibits apoptotic cell death induced in Rat-1a cells by multifunctional cytokine (Klefstrom et al. 1994) and in M1 myeloid leukemia cells by TGFβ (Selvakumaran et al. 1994).

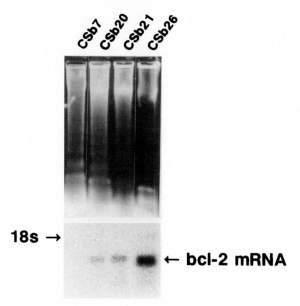

18s →

← bcl-2 mRNA

Fig. 15. Inhibition of s-Myc-induced apoptotic cell death by *bcl-2* gene expression. *Upper panel* DNA fragmentation in s-*myc* transfectants introduced with the *bcl-2* genes. C6smyc4, C6smyc4-b20, C6smyc4-b21, and C6smyc4-b26 cells contain the 0, 3, 5, and 9 copies of the exogenously transfected *bcl-2* gene, respectively. *Lower panel* The expression level of bcl-2 mRNA (1100 nts) in each transfectant)

These findings led to examining whether s-Myc-mediated apoptosis is suppressed by *bcl-2* gene expression. For this study, the human *bcl-2* cDNA was subcloned in an episomal expression vector pCEP4 containing the cytomegalovirus promoter and the hygromycin-resistant gene, and transfected into glioma cells. As shown in Fig. 15, DNA fragmentation produced by zinc sulfate-induced s-Myc expression was inhibited proportionally with the increased expression of the *bcl-2* gene. These results clearly indicate that s-Myc-mediated signal transduction to induce apoptosis is also inhibited by Bcl-2 expression as well as that induced by c-Myc expression.

In addition to Bcl-2, IL-6 and retinoic acid can inhibit c-*myc*-induced apoptotic cell death. Interleukin 6 induces differentiation of the murine hematopoietic cell line Y6 into macrophages and apoptosis of these cells, and in this system, IL-6-induced macrophage differentiation and apoptotic cell death is preceded by downregulation of the c-*myc* gene. This c-*myc* downregulation followed by induction of apoptosis is prevented by retinoic acid which is known to inhibit neoplasia and inflammation and induce differentiation (Oritani et al. 1992). Yonish-Rouach et al. (1991, 1993) reported that IL6 induces differentiation of the murine myeloid leukaemic line M1 and also markedly inhibits death of these cells induced by repression of the c-myc gene due to deregulated wt p53 expression.

5 Conclusions

The *c-myc* and s-*myc* genes are regulatory genes that differ in cell type specificity for influencing susceptibility to apoptosis: c-Myc overexpression induces apoptosis in several types of cells, including cultured immortalized fibroblast Rat-1 and NIH3T3 cells, when those cells are deprived of serum growth factors, whereas s-Myc overexpression selectively triggers apoptosis of neural tumor cells cultivated in the presence of growth factors. Even et al. (1992) speculated that on withdrawal of serum growth factors, c-*myc* gene expression is downregulated and the cells revert to a growth-arrested state in which they may remain viable. In these growth-arrested cells, forced expression of the c-*myc* gene causes apoptotic death of a substantial number of the cells. A similar stimulative effect of c-Myc expression on apoptosis induction was observed in IL-3-deprived myeloid cells modified to express the c-*myc* gene constitutively (Askew et al. 1991; Cleveland et al. 1994; Malde and Collins 1994; Marvel et al. 1994). IL-3-deprivation decreased c-Myc expression in the myeloid cells and arrested the cell cycle in the G1 phase. In these cells, constitutive expression of c-Myc enhanced susceptibility to apoptotic cell death. Taken together with other evidence (Klefstrom et al. 1994), these findings reveal that cell cycle arrest induced by deprivation of a serum growth factor or cytokine might play a role in initiating apoptosis.

Recently, we have found that constitutive expression of an exogenously transfected s-*myc* gene inhibited cellular progression from the G1 to S phase, although s-Myc expression itself had no effect on expression of the endogenous c-*myc* gene (Asai et al. 1994b). Therefore, s-Myc may be able to induce apoptosis in rat and human glioma cells even when the cells are not deprived of serum growth factors. Nevertheless, s-Myc-mediated apoptosis is blocked by *bcl-2* gene expression as c-Myc-mediated apoptosis is. The s-Myc protein contains all the domain structures conserved in Myc proteins such as c-Myc and N-Myc except the CKII sites in the internal acidic domain (Sugiyama et al. 1989). The conservation of these functional domains between c-Myc and s-Myc implies that s-Myc may have biological functions similar to those of c-Myc. If this is so, s-Myc may be involved in the similar stage of apoptotic pathway as c-Myc to regulate the transcription expression of the gene(s) involved in apoptosis. However, unlike c-Myc, s-Myc can induce apoptosis in human and rat glioma cells expressing only mutated p53 (Asai et al. 1994a,b). Recently, Hermeking and Eick (1994) demonstrated that activation of c-Myc in quiescent mouse fibroblast cells leads to an accumulation of wt p53, resulting in apoptosis induction and prevention of cell cycle reentry. Similar evidence indicating that expression of wt p53 is required for c-Myc-mediated apoptosis has been observed by Steinman et al. (1994) and Wagner et al. (1994). Furthermore, as reported by Facchini et al. (1994), c-Myc overexpression can induce both cell proliferation and apoptosis in nontransformed cells, but c-*myc* activation is commonly tolerated in many tumors such as the tumorigenic L929 cell line. These findings strongly suggest that s-Myc may induce apoptosis in glioma cells through a process differing from that of c-Myc-mediated apoptosis.

For induction of apoptosis, the association of the Myc protein with a partner protein Max is required (Amati et al. 1993; Bissonnette et al. 1994). The Max protein is a stable bHLH-Zip protein and its expression level is almost constant during the cell cycle (Blackwood et al. 1991). We recently found that the *max* gene is transcribed at a moderate level in malignant glioma cells, and that Max associated in vitro with s-Myc as well as c-Myc and N-Myc (unpubl.). Therefore, cell-type specifity of apoptosis may not be determined simply by the extent of s-Myc-Max complex formation in cells.

Recently, Gu et al. (1994) demonstrated that an N-terminal domain present in the c-Myc protein mediates binding to the p107 Rb-related protein, whose binding suppresses the transactivation activity of the c-Myc. Moreover, Shrivastava et al. (1993) found that c-Myc associates with zinc finger protein Yin-Yang-1 (YY-1) and causes inhibition of the activity of YY1. For association with YY1, the regions of c-Myc containing a CKII site, a nuclear localization signal, and a nonspecific DNA binding signal are required. These observations suggest that a second factor similar to p107 and YY1, which may be specific to glioma cells and associate selectively with s-Myc, may be required for the induction of cell-type-specific apoptosis. If this is so, it should be interesting to characterize this second factor to understand the mechanism of induction of cell-type-specific apoptosis.

Acknowledgments. We thank Drs. A. Sugiyama, Y. Miyagi, K. Mishima, H. Kanemitsu, and Y. Nagashima for their collaboration. This work was partly supported by a Grant-in-Aid from the Ministry of Health and Welfare of Japan for a Comprehensive 10-Year Strategy for Cancer Control and by a Grant from the Ministry of Health and Welfare and the Ministry of Education, Science, and Culture in Japan.

References

Aeschlimann D, Wetterwald A, Fleisch H, Paulsson M (1993) Expression of tissue transglutaminase in skeletal tissues correlates with events of terminal differentiation of chondrocytes. J Cell Biol 120: 1461–1470

Alnemri ES, Fernandes TF, Haldar S, Croce CM, Litwack G (1992) Involvement of *bcl-2* in glucocorticoid-induced apoptosis of human pre-B-leukemias. Cancer Res 52: 491–495

Amati B, Littlewood TD, Evan GI, Land H (1993) The c-Myc protein induces cell cycle progression and apoptosis through dimerization with Max. EMBO J 12: 5083–5087

Arends MJ, Wyllie AH (1991) Apoptosis: mechanisms and roles in pathology. Int Rev Exp Pathol 32: 223–254

Asai A, Miyagi Y, Sugiyama A, Gamanuma M, Hong SI, Takamoto S, Nomura K, Matsutani M, Takakura K, Kuchino Y (1994a) Negative effects of wild-type p53 and s-Myc on cellular growth and tumorigenicity of glioma cells: implication of the tumor suppressor genes for therapy. J Neurooncol 19: 259–268

Asai A, Miyagi Y, Sugiyama A, Nagashima Y, Kanemitsu H, Obinata M, Mishima K, Kuchino Y (1994b) The s-Myc protein having the ability to induce apoptosis is selectively expressed in rat embryo chondrocytes. Oncogene 9: 2345–2352

Askew DS, Ashmun RA, Simmons BC, Cleveland Jl (1991) Constitutive c-*myc* expression in an IL-3-dependent myeloid cell line suppresses cell cycle arrest and accelerates apoptosis. Oncogene 6: 1915–1922

Bertrand R, Sarang M, Jenkin J, Kerrigan D, Pommier Y (1991) Differential induction of secondary DNA fragmentation by topoisomerase II inhibitors in human tumor cell lines with amplified c-*myc* expression. Cancer Res 51: 6280–6285

Bissonnette RP, Echeverri F, Mahboubi A, Green DR (1992) Apoptotic cell death induced by c-myc is inhibited by *bcl-2*. Nature 359: 552–554

Bissonnette RP, MaGahon A, Mahboubi A, Green DR (1994) Functional Myc-Max heterodimer is required for activation-induced apoptosis in T cell hybridomas. J Exp Med 180: 2413–2418

Blackwood EM, Eisenman RN (1991) Max: a helix-loop-helix zipper protein that forms a sequence-specific DNA binding complex with Myc. Science 251: 1211–1217

Blackwood EM, Luscher B, Kretzner L, Eisenman RN (1991) The Myc: Max protein complex and cell growth regulation. Cold Spring Harbor Symp Quant Biol LVI: 109–117

Blackwood EM, Luscher B, Eisenman RN (1992) Myc and Max associate in vivo. Genes Dev 6: 71–80

Clarke AR, Purdie CA, Harrison DJ, Morris RG, Bird CC, Hooper ML, Wyllie AH (1993) Thymocyte apoptosis induced by p53-dependent and independent pathways. Nature 362: 849–851

Cleveland JL, Troppmair J, Packham G, Askew DS, Lloyd P, Gonzalez-Garcia M, Nunez G, Ihle JN, Rapp UR (1994) v-raf suppresses apoptosis and promotes growth of interleukin-3-dependent myeloid cells. Oncogene 9: 2217–2226

Davidoff AN, Mendelow BV (1993) Substituted purines elicit differential cytokinetic, molecular and phenotypic effects in HL-60 cells. Exp Hematol 21: 456–601

Duke RC, Cohen JJ (1986) IL2 addiction. Limphokine Res 5: 289–299

Ellis RE, Yuan J, Horvitz HR (1991) Mechanisms and functions of cell death. Annu Rev Cell Biol 7: 663–698

Evan GI, Wyllie AH, Gilbert CS, Littlewood TD, Land H, Brooks M, Waters CM, Penn LZ, Hancock DC (1992) Induction of apoptosis in fibroblasts by c-myc protein. Cell 69: 119–128

Facchini LM, Chen S, Penn LJ (1994) Dysfunction of the Myc-induced apoptosis mechanism accompanies c-myc activation in the tumorigenic L929 cell line. Cell Growth Differ 5: 637–646

Fanidi A, Harrington EA, Evan GI (1992) Cooperative interaction between c-*myc* and *bcl-2* proto-oncogenes. Nature 359: 554–556

Farquharson C, Hesketh JE, Loveridge N (1992) The proto-oncogene c-myc is involved in cell differentiation as well as cell proliferation: studies on growth plate chondrocytes in situ. J Cell Physiol 152: 135–144

Fesus L, Tarcsa E, Kedei N, Autuori F, Piacentini M (1991) Degradation of cells dying by apoptosis leads to accumulation of epsilon (gamma-glutamyl) lysine isodipeptide in culture fluid and blood. FEBS Lett 284: 109–112

Fourel G, Trepo C, Bougueleret L, Henglein B, Ponzetto A, Tiollais P, Buendia MA (1990) Frequent activation of N-*myc* genes by hepadnavirus insertion in woodchuck liver tumors. Nature 347: 294–298

Gorman CM, Moffat LF, Howard BH (1982) Recombinant genomes which express chloramphenicol acetyltransferase in mammalian cells. Mol Cell Biol 2: 1044–1051

Gu W, Bhatia K, Magrath IT, Dang CV, Dalla-Favera R (1994) Binding and suppression of the Myc transcriptional activation domain by p107. Science 264: 251–254

Hermeking H, Eick D (1994) Mediation of c-Myc-induced apoptosis by p53. Science 265: 2091–2093

Hoang AT, Cohen KJ, Barrett JF, Bergstorm DA, Dang CV (1994) Participation of cyclin A in Myc-induced apoptosis. Proc Natl Acad Sci USA 91: 6875–6879

Hockenbery D, Nunez G, Milliman C, Schreiber RD, Korsmeyer SJ (1990) Bcl-2 is an inner mitochondrial membrane protein that blocks programmed cell death. Nature 348: 334–336

Ivanov X, Mladenov Z, Nedyalkov S, Todorov TG, Yakimov M (1964) Experimental investigations into avain leukoses. V. Transmission, haematology and morphology of avian myelocytomatosis. Bull Inst Pathol Comp Anim 10: 5–38

Johnson RS, Spiegelman BM, Papaioannou V (1992) Pleiotropic effects of a null mutation in the c-*fos* proto-oncogene. Cell 71: 577–586

Kitanaka C, Sugiyama A, Kanazu S, Miyagi Y, Mishima K, Asai A, Kuchino Y (1995) s-Myc acts as a transcriptional activator and its sequence-specific DNA binding is required for induction of programmed cell death in glioma cells. Cell Death Differ 2: 123–131

Klefstrom J, Vastrik I, Saksela E, Valle J, Eilers M, Alitalo K (1994) c-Myc induces cellular susceptibility to the cytotoxic action of TNF-alpha. EMBO J 13: 5442–5450

Kretzner L, Blackwood EM, Eisenman RN (1992) Myc and Max proteins possess distinct transcriptional activities. Nature 359: 426–429

Kuchino Y, Sugiyama A, Miyagi Y (1990) s-Myc, a new member of the myc family has tumor suppressing activity. Yokohama Med Bull 41: 71–74

Lotem J, Sachs L (1993) Regulation by *bcl-2*, c-*myc* and *p53* of susceptibility to induction of apoptosis by heat shock and cancer chemotherapy compounds in differentiation-competent and -defective myeloid leukemic cells. Cell Growth Differ 4: 41–47

Lowe SW, Schmitt EM, Smith SW, Osborne BA, Jacks T (1993) p53 is required for radiation-induced apoptosis in mouse thymocytes. Nature 362: 847–849

Luscher B, Eisenman RN (1990) c-myc and c-myb protein degradation: effect of metabolic inhibitors and heat shock. Mol Cell Biol 8: 2504–2512

Malde P, Collins MK (1994) Disregulation of Myc expression in murine bone marrow cells results in an inability to proliferate in sub-optimal growth factor and an increased sensitivity to DNA damage. Int Immunol 6: 1169–1176

Marvel J, Perkins GR, Lopez RA, Collins MK (1994) Growth factor starvation of *bcl-2* over-expressing murine bone marrow cells induced refractoriness to IL-3 stimulation of proliferation. Oncogene 9: 1117–1122

Marx J (1993) Cell death studies yield cancer clues. Science 259: 760–761

Meek DW, Street AJ (1992) Nuclear protein phosphorylation and growth control. Biochem J 287: 1–15

Oren M (1992) The involvement of oncogenes and tumor suppressor genes in the control of apoptosis. Cancer Meta Rev 11: 141–148

Oritani K, Kaisho T, Nakajima K, Hirano T (1992) Retinoic acid inhibits interleukin-6-induced macrophage differentiation and apoptosis in murine hematopoietic cell line, Y6. Blood 80: 2298–2305

Owens GP, cohen JJ (1992) Identification of genes involved in programmed cell death. Cancer Meta Rev 11: 149–156

Packham G, Cleveland JL (1994) Ornithine decarboxylase is a mediator of c-Myc-induced apoptosis. Mol Cell Biol 14: 5741–5747

Pallavicini MG, Rosette C, Reitsma M, Deteresa PS, Gray LW (1990) Relationship of c-*myc* gene copy number and gene expression: cellular effects of elevated c-myc protein. J Cell Physiol 143: 372–380

Parkin N, Darveau A, Nicholson R, Sonenberg N (1988) *cis*-Acting translational effects of the 5' noncoding region of c-myc mRNA. Mol Cell Biol 8: 2875–2883

Penn LJZ, Laufer EM, Land H (1990) c-myc: evidence for multiple regulatory functions. Semin Cancer Biol 1: 68–87

Prendergast GC, Lawe D, Ziff EB (1991) Association of Myn, the murine homolog of Max, with c-Myc stimulates methylation-sensitive DNA binding and Ras cotransformation. Cell 65: 395–407

Raff MC. (1992) Social controls on cell survival and cell death. Nature 356: 397–400

Reynolds JE, Yang T, Qian L, Jenkinson JD, Zhou P, Eastman A, Craig RW (1994) Mcl-1, a member of the Bcl-2 family, delays apoptosis induced by c-Myc overexpression in Chinese hamster ovary cells. Cancer Res 54: 6348–6352

Schwab M (1988) The *myc*-box oncogenes. In: Reddy EP, Skalka AM, Curran T (eds) The oncogene handbook. Elsevier, New York, pp 381–391

Schwartz LM, Smith SW, Jones MEE, Osborne BA (1993) Do all programmed cell death occur via apoptosis? Proc Natl Acad Sci USA 90: 980–984

Selvakumaran M, Lin HK, Sjin RT, Reed JC, Liebermann DA, Hoffman B (1994) The novel primary response gene MyD118 and the protooncogenes *myb*, *myc*, and *bcl-2* modulate trans-forming growth factor beta 1-induced apoptosis of myeloid leukemia cells. Mol Cell Biol 14: 2352–2360

Shaw P, Bovey R, Tardy S, Sahli R, Sordat B, Costa J (1992) Induction of apoptosis by wild-type p53 in a human colon tumor-derived cell line. Proc Natl Acad Sci USA 89: 4495–4499

Sheiness D, Fanshier L, Bishop JM (1978) Identification of nucleotide sequences which may encode the oncogenic capacity of avian retrovirus MC29. J Virol 28: 600–610

Shi Y, Glynn JM, Guilbert LJ, Cotter TG, Bissonnette RP, Green DR (1992) Role for c-myc in activation-induced apoptotic cell death in T cell hybridomas. Science 257: 212–214

Shrivastava A, Saleque S, Kalpana GV, Artandi S, Goff SP, Calame K (1993) Inhibition of transcriptional regulator Yin-Yang-1 by association with c-Myc. Science 262: 1889–1892

Smeyne RJ, Vendrell M, Hayward M, Baker SJ, Miao GG, Schilling K, Robertson LM, Curran T, Morgan JI (1993) Continuous c-*fos* expression precedes programmed cell death in vivo. Nature 363: 166–169

Steinman RA, Hoffman B, Iro A, Guillouf C, Liebermann DA, el-Houseini ME (1994) Induction of p21 (WAF-1/CIP1) during differentiation. Oncogene 9: 3389–3396

Stone J, Lange T, Ramsay G, Jakobovits E, Bishop JM, Varmus HE, Lee W (1987) Definition of regions in human c-*myc* that are involved in transformation and nuclear localization. Mol Cell Biol 7: 1697–1709

Strasser A, Harris AW, Bath ML, Cory S (1990) Novel primitive lymphoid tumors induced in transgenic mice by cooperation between *myc* and *bcl-2*. Nature 348: 331–333

Sugiyama A, Kume A, Nemoto K, Lee SY, Asami Y, Nemoto F, Nishimura S, Kuchino Y (1989) Isolation and characterization of s-*myc*, a member of rat *myc* gene family. Proc Natl Acad Sci USA 86: 9144–9189

Tepper CG, Studzinski GP (1992) Teniposide induces nuclear but not mitochondrial DNA degradation. Cancer Res 52: 3384–3390

Thulasi R, Harbour DV, Thompson EB (1993) Suppression of c-*myc* is a critical step in glucocorticoid-induced human leukemic cell lysis. J Biol Chem 268: 18306–18312

Tsujimoto Y, Finger L, Yunis J, Nowell P, Croce C (1984) Involvement of the *bcl-2* gene in human follicular lymphoma. Science 226: 1097–1099

Vaux DL (1993) Toward an understanding of the molecular mechanisms of physiological cell death. Proc Natl Acad Sci USA 90: 786–789

Vaux DL, Weissman IL (1993) Neither macromolecular synthesis nor myc is required for cell death via the mechanism that can be controlled by Bcl-2. Mol Cell Biol 13: 7000–7005

Vaux DL, Cory S, Adams JM (1988) *Bcl-2* gene promotes haemopoietic cell survival and cooperates with c-*myc* to immortalize pre-B cells. Nature 335: 440–442

Wagner AJ, Small MB, Hay N (1993) Myc-mediated apoptosis is blocked by ectopic expression of Bcl-2. Mol Cell Biol 13: 2432–2440

Wagner AJ, Kokontis JM, Hay N (1994) Myc-mediated apoptosis requires wild-type p53 in a manner independent of cell cycle arrest and the ability of p53 to induce p21waf1/cip1. Gene Dev 8: 2817–2830

Williams GT (1991) Programmed cell death: apoptosis and oncogenesis. Cell 65: 1097–1098

Wu FY, Chang NT, Chen WJ, Juan CC (1993) Vitamin K3-induced cell cycle arrest and apoptotic cell death are accompanied by altered expression of c-*fos* and c-*myc* in nasopharyngeal carcinoma cells. Oncogene 8: 2237–2244

Wyllie AH (1992) Apoptosis and the regulation of cell numbers in normal and neoplastic tissues: an overview. Cancer Meta Rev 11: 95–103

Wurm FM, Gwinn KA, Kingston RE (1986) Inducible overproduction of the mouse c-myc protein in mammalian cells. Proc Natl Acad Sci USA 83: 5414–5418

Yonish-Rouach E, Resnitzky D, Lotem J, Sachs L, Kimchi A, Oren M (1991) Wild-type p53 induces apoptosis of myeloid leukaemic cells that is inhibited by interleukin-6. Nature 352: 345–347

Yonish-Rouach E, Grunwald D, Wilder S, Kimchi A, May E, Lawrence JJ, May P, Oren M (1993) p53-mediated cell death: relationship to cell cycle control. Mol Cell Biol 13: 1415–1423

Clusterin: A Role in Cell Survival in the Face of Apoptosis?

C. Koch-Brandt[1] and C. Morgans[2]

Abstract

Clusterin is a multifunctional glycoprotein complex found in virtually all body fluids and on the surface of cells lining body cavities. Demonstrated and proposed functions include the transport of lipoproteins, the inhibition of complement-mediated cell lysis and the modulation of cell-cell interactions. On the basis of its elevated expression in apoptotic tissues, it was originally proposed that the protein might be casually involved in apoptosis. Here, we discuss the recent data that, in contrast to the earlier notion, suggest that clusterin expression is not enhanced, but rather is down-regulated in the cells undergoing apoptosis and that its expression in the apoptotic tissue is restricted to the vital neighboring cells. These results led to the proposal that rather than being a cell death gene, clusterin is a cell survival gene, exerting a protective function on the surving bystander cells.

1 Introduction

Apoptosis (programmed cell death) is a normal process that allows tissue remodelling during development and by counteracting proliferation ensures the maintenance of constant cell numbers in differentiated tissues (Raff 1992). Perturbation of this equilibrium by suppression of apoptosis can lead to oncogenesis, while treatment of malignant tumors by radiation and chemotherapy has been proposed to trigger the induction of apoptosis (Bursch et al. 1985; Wyllie 1985; Williams 1991; Smith et al. 1994). Finally, apoptosis is thought to be a mechanism of damaged cell elimination during tissue injury (Bursch et al. 1992a).

There are characteristic morphological and biochemical events that have been used to define apoptosis and to distinguish it from necrosis (Kerr and Harmon 1991). The histological hallmarks of apoptosis in chronological order of appearance are the condensation of nuclear chromatin, the fragmentation of the nucleus, the convolution of the cell membranes, and finally, the fragmentation of the cell into apoptotic bodies which are phagocytosed and degraded by neighboring cells

[1]Institut für Biochemie, Johannes Gutenberg-Universität, Becherweg 30, 55099, Mainz, Germany
[2]Max Planck Institut für Hirnforschung, Abteilung Neurochemie, Deutschordenstr. 46, 60528 Frankfurt/Main, Germany

and macrophages. Biochemically, apoptosis has best been characterized in lymphocytes. It is associated with the endonucleolytic breakdown of the chromosomal DNA into oligonucleosomes by the action of Ca^{2+}/Mg^{2+}-dependent DNase(s), which gives rise to a characteristic ladder-like pattern upon agarose gel electrophoresis (Wyllie 1980; Kyprianou et al. 1988; Peitsch et al. 1994). Recent evidence suggests that it involves at least two distinct steps (Filipski et al. 1990; Brown et al. 1993; Oberhammer et al. 1993; Tanuma and Shiokawa, this Vol.). In contrast to necrosis, in which inflammatory events tend to spread within the tissue, apoptosis results in the death of the affected cells with minimal damage to adjacent cells (Williams et al. 1992; Collins and Rivas 1993; Martin 1993).

There is now a broad consensus that apoptosis is a gene-directed process. Although it had been realized long ago that apoptosis is dependent on ongoing RNA and protein synthesis (Wadewitz and Lockshin 1988; Buttyan 1991; Collins and Rivas 1993; Tenniswood et al. 1994), only recently have specific genes been identified that appear to be directly involved in the control of apoptosis in *Caenorhabditis elegans*, *Drosophila*, and mammals (Ellis et al. 1991; Williams and Smith 1993; Osborne and Schwartz 1994; White et al. 1994). In mammalian cells, a concerted action of genes, including the tumor suppressor gene p53 and the proto-oncogene c-*myc*, has been shown to induce apoptosis, while the proto-oncogene bcl-2 and related genes act to suppress apoptosis (Williams and Smith 1993; Bissonnette et al. 1994; Furuya et al. 1994; Smith et al. 1994; see Kuchina and Asai, Tsujimoto, this Vol.). Furthermore, some viruses, such as HIV, leukemia viruses, and adenovirus, use induction or suppression of host cell apoptosis as an obligatory step in their life cycle (Gougeon and Montagnier 1994; Hovanessian 1994; White and Gooding 1994; Rojko et al., this Vol.). Since apoptosis is subject to control from external factors such as hormones, growth factors, and as yet unidentified components, several second messenger systems – the best characterized being the APO-1/FAS-dependent signal transduction – have been implicated in the control of apoptosis (Lee et al. 1993; Nagata 1994; Nagata, this Vol.).

Apart from these apparently causal genes, the ordered series of events characteristic of apoptosis depends on the action of additional gene products that are responsible for the characteristic morphological and biochemical changes in the apoptotic cell or are involved in the protection of the neighboring cells. These proteins, although not directly involved in the induction of apoptosis, constitute an important activity within the entire process and are potentially valuable as cellular markers of apoptosis. The search for proteins overexpressed in apoptotic tissue has led to the identification of clusterin (Apolipoprotein J, TRPM-2, and other names), tissue transglutaminase poly(ADP-ribose)polymerase, hsp27 and hsp70, the Yb1 subunit of glutathione S-transferase, cathepsin D, proteins called RVP.1, RP-2, and RP-8, tissue and urokinase-type plasminogen activators, and glucocorticoid and cAMP-induced genes from T lymphocytes (Tenniswood et al. 1994). In particular, clusterin has surfaced repeatedly as an overexpressed protein in model systems of apoptosis and injury, and showed promise early on to be a possible marker of apoptosis. Recent, more detailed investigations into the association between clusterin expression and apoptosis, however, have revealed that, although intriguing, this association needs careful analysis and interpretation.

This chapter will focus on these latest findings about the association and possible role of clusterin in several model systems of apoptosis.

2 Clusterin – a Widely Expressed Multifunctional Protein

Clusterin is a heterodimeric 80-kDa glycoprotein found in virtually all body fluids and on the surface of cells lining body cavities (Table 1; Jenne and Tschopp 1992). Originally, the protein was identified in the male reproductive tract. It was isolated from ram rete testis fluid on the basis of its aggregating activity, in vitro, on erythrocytes and Sertoli cells (Blaschuk et al. 1983; Fritz et al. 1983). The protein was further identified as a major glycoprotein secreted by rat Sertoli cells in culture and called SGP-2 (sulfated glycoprotein-2; Griswold et al. 1986; Collard and Griswold 1987), and as one of the gene products whose mRNA was specifically upregulated in the prostate after castration (testosterone-repressed prostate message-2, TRPM-2; Montpetit et al. 1986). The comparison of the cDNA-derived sequences of SGP-2 and TRPM-2 with the amino acid sequence of purified sheep clusterin in conjunction with Southern blot analysis revealed that the three proteins were species homologues.

More recently, additional species homologues have been identified in a broad spectrum of biological contexts, suggesting that the protein is multifunctional. The human form was isolated from serum as a constituent of the soluble, nonlytic terminal complement complex (CLI, Sp 40, 40) and as a component of the high-density lipoprotein (HDL) fraction (NA1/NA2, apolipoprotein J; Murphy et al. 1988; Jenne and Tschopp 1989; Kirszbaum et al. 1989; De Silva et al. 1990). The

Table 1. Clusterin and its species homologues

Name	Characteristic	Species	Organ
Clusterin	Protein from testis with cell-aggregating activity	Sheep	Testis
TRPM-2	mRNA-induced in rat prostate after castration	Rat	Prostate
SGP-2	Protein secreted from Sertoli cells	Rat	Testis
CLI	Secretory protein expressed in cultured smooth muscle cells	Pig	Thoracic aorta
CLI, SP40, 40	Inhibitor of complement-mediated lysis	Human	Serun
Apo J, NA1/NA2	Apolipoprotein	Human	Serum
T64	mRNA induced in neuroretinal cells by RSV	Quail	Neuroretina
pTB16	mRNA overexpressed in glioma and epileptic	Human	Brain
pADHC-9	foci mRNA increased in Alzheimer's disease	Human	Brain
HISL-19	Immunocytochemical marker of neuroendocrine cells	Human	Brain
gp-III	Constituent of the secretory granula in chromaffin cells	Bovine	Adrenal medulla
gp 80	Apically secreted protein in MDCK cells	Dog	Kidney

porcine homologue was shown to be expressed in cultured aortic smooth muscle cells (Diemer et al. 1992). The protein (referred to as gp80 in this system) has also been studied as a marker for vectorial transport to the apical cell surface in the canine kidney-derived epithelial cell line MDCK (Urban et al. 1987; Hartmann et al. 1991), as a constituent of the chromaffin granules of bovine adrenal medulla cells (GP III) and as an immunocytochemical marker (HISL 19) of human neuroendocrine cells (Palmer and Christie 1990; Huttner et al. 1991; Laslop et al. 1993). In the human nervous system, the mRNA encoding the protein has been shown to be elevated in the Alzheimer's disease (pADHC-9), in gliomas and epileptic foci (pTB16; Duguid et al. 1989; Danik et al. 1991). In quail, the mRNA has been found to be induced by RSV infection in neuroretinal cells (Michel et al. 1989).

The biogenesis of the protein has been studied extensively. It is synthesized as a 65-kDa single chain precursor protein in its high mannose ER-associated form. As the protein moves along the exocytic pathway, the N-linked carbohydrate structures mature and the protein undergoes extensive sulfation, predominantly, if not exclusively, at the carbohydrate moieties. The precursor is proteolytically cleaved into two subunits which are released from the cell as a disulfide-linked heterodimeric complex (Urban et al. 1987). Secretion has been shown to be under strict temporal and topological control. Luminal secretion has been demonstrated in the kidney-derived epithelial cell line MDCK (Urban et al. 1987), in thyroid epithelial cells (Koch-Brandt, unpubl.), after gene transfer in intestinal cells (Appel and Koch-Brandt 1994), and presumably also occurs in the epithelial cells of the prostate, testes, epididymis, and mammary glands. Hepatocytes which lack a direct exocytic route to the apical (bile canicular) surface release clusterin, (presumably after intracellular association with apolipoprotein A – I) at the basolateral (sinusoidal) surface into the serum (Jenne et al. 1991). In cells featuring a regulated exocytic pathway, the release of clusterin is dependent upon the appropriate exogenous stimulus. Secretion of clusterin (apolipoprotein J) from megakaryocytes occurs upon thrombin induced-platelet activation (Tschopp et al. 1993; Witte et al. 1993). In the pheochromocytoma cell line PC12, depolarization-induced secretion has been demonstrated (Appel and Koch-Brandt, unpubl.).

The primary structure of the protein, as deduced from the cDNA sequence, revealed the presence of five strictly symmetrically arranged cysteines on each of the subunits which participate in the intersubunit disulfide bridges (Cheng et al. 1988; Choi-Miura et al. 1992; Kirszbaum et al. 1992). Each subunit carries two regions predicted to form amphipatic α-helices and, depending upon the species, 2–4 N-glycosylation sites. On the N-terminal subunit, regions with homology to known Ca^{2+}-, heparin- and dinucleotide-binding sites as well as the most conserved region are found (Fig. 1).

The clusterin (TRPM-2) gene was first cloned from rat (Wong et al. 1993). Subsequently, the human, mouse, bovine, and canine genes were characterized and found to be highly homologous in structure. The gene exists as a single copy located on human chromosome 8 and on mouse chromosome 14 (Purello et al.

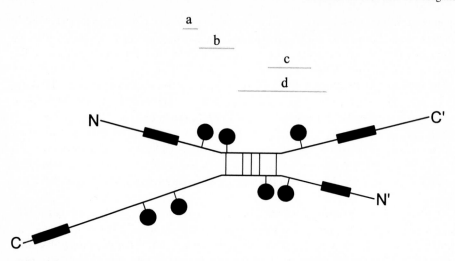

Fig. 1. Schematic diagram of the proposed clusterin structure. N and C are the amino- and carboxytermini of the precursor protein, N' and C' are the termini produced by proteolytic processing. *Filled circles* represent N-linked carbohydrate moieties; *Filled bars* indicate potential amphipatic helices; *lines between the subunits* symbolize disulfide bridges; *bars above the drawing* designate potential binding sites for heparin (*a*), Ca^{2+} (*b*), dinucleotides (*c*), and the most conserved region (*d*)

1991; Jordan-Starck et al. 1992). The conservation of the nucleotide sequence within the coding region ranges from 82% (human-dog) to 49% (human-quail) (Jenne and Tschopp 1992). Homologous in *Drosophila* and yeast have been looked for using heterologues screening with oligonucleotides derived from a highly conserved region of the cDNA, but have so far not been detected (Koch-Brandt, unpubl.).

The expression of the gene has been shown to be under negative control by cAMP (Leydig tumor cells), steroid hormones (testis, prostate, and mammary carcinoma cells) and glucocorticoids (kidney epithelial cells) and induced by stimuli elicited from tissue injury. These data indicate that the gene is under complex regulation and that multiple signal transduction pathways participate in the modulation of its activity (Sensibar et al. 1991; Pignatoro et al. 1992).

During embryogenesis, the expression of clusterin gene has been shown to be under strict temporal and spatial control. In situ hybridization studies revealed that in the developing mouse clusterin is expressed nearly exclusively in developing epithelia (French et al. 1993). This was especially evident in tissues such as skin, duodenum, and tooth, where proliferative and differentiating regions are easy to distinguish. In lung and kidney, clusterin was expressed in the differentiating bronchial and tubular epithelial cells. In the developing kidney, clusterin expression was detected 14 days after conception. At this stage, organogenesis of the kidney, which includes the de novo formation of kidney epithelium from unpolarized mesenchyme, has already given rise to a small number of nephric units containing polarized tubular epithelium (Hartmann et al. 1991). On the basis of these results, it has been suggested that the protein might be involved in the differentiation and morphogenesis of epithelia. The precise function(s) of clusterin in these processes remain to be defined, but they could include the modulation of cell-cell and/or cell-matrix interactions.

Although the protein has been assigned different functions in the different systems, all of these functions could be ascribed to the propensity of clusterin to bind to hydrophobic domains on macromolecules. First, it has been shown to bind to hydrophobic domains in nascent C5b-7, C5b-8, or C5b-9 complement complexes and to inhibit their insertion into the membrane (Jenne and Tschopp 1992). Second, its association with HDL particles and its dissociation from these lipoprotein particles by nonionic detergents suggest a direct interaction with lipids (Jenne et al. 1991). Third, the modulation of cell-cell interactions by clusterin appears to involve lipid binding since its cell aggregating activity in vitro can be inhibitied by long-chain fatty acyl derivatives of carnitine (Fritz and Burdzy 1989). Other postulated functions of the protein, such as a proposed role in tissue remodeling and repair by binding to and subsequent removal of membrane fragments and cell debris, are also based on the involvement of clusterin in hydrophobic interactions (Jenne et al. 1991).

3   Clusterin Gene Expression in Apoptotic Epithelial Tissues

There is broad evidence for role of clusterin in epitelial apoptosis, oncogenesis, and injury. It has been shown that clusterin mRNA is induced in epithelial tumors regressing during hormonal treatment or chemotherapy (Rennie et al. 1988; Buttyan et al. 1989; Kyprianou et al. 1990; Kyprianou et al. 1991). Clusterin (SGP-2) mRNA has been shown to be elevated in the epithelial components of rat prostate and seminal vesicle carcinomas (Kadomatsu et al. 1993) and in the prostate regressing after castration (Buttyan 1991). In clear cell tumors of the kidney, clusterin (gp80) mRNA levels are three-to four fold elevated when compared to the levels in the normal kidney tissue of the same patient (Parczyk et al. 1994). Chronic obstruction of the ureter leads to hydronephrosis and highly increased expression of clusterin in the epithelial cells of the collecting ducts and distal tubules (Buttyan et al. 1989; Sawczuk et al. 1989), which are the sites of cellular destruction and apoptosis. The aminoglycoside-induced enhancement of clusterin expression in the kidney has been shown to correlate with the development of acute tubular destruction. (Brandyk et al. 1990). Furthermore, one of the first descriptions of clusterin was its isolation from the immune deposits of patients with membranous glomerulonephritis (Murphy et al. 1988). Correlations between the occurrence of apoptosis and changes in clusterin gene expression have been investigated in detail in liver, in hormone-dependent breast cancer and in the regressing prostate after castration.

In liver, apoptosis has been studied extensively in normal, injured, and (pre-) neoplastic tissue, and it has been suggested that apoptosis may be one of the primary mechanisms for the prevention of tumor formation (Bursch et al. 1992b).

Clusterin gene expression has been analyzed by in situ hybridization and Northern blot analysis in response to hepatomitogen administration and withdrawal. While in control livers all hepatocytes express low levels of clusterin mRNA, the clusterin levels are increased both during liver growth and regression. The enhanced expression is not confined to the dying cells but is detected in the entire hepatocyte population. On the basis of these data, it has been proposed that clusterin may assist in the maintenance of membrane integrity which is crucial during both mitosis and apoptosis. Furthermore, enhanced clusterin expression was also induced by necrogenic doses of carbon tetrachloride, suggesting that the role of clusterin in membrane remodeling is a general one and not specifically associated with either mitosis or apoptosis. This notion is supported by the low level expression in control hepatocytes which may generate clusterin in amounts just sufficient to support the continuous membrane remodeling during normal hepatocyte differentiation and turnover (Bursch et al. 1995).

In hormone-dependent breast cancer, hormonal ablation leads to massive apoptosis and tumor regression (Davidson and Lippman 1988). Analyses of clusterin gene expression during the regression phase have been reported for the androgen-dependent Shionogi mouse mammary carcinoma and for the estrogen-dependent MCF-7 human breast cancer. When Shionogi carcinoma cells were injected into adult male mice and grown for 15 days to yield tumors of 3–5 g by weight, castration of the animals led to tumor regression starting 72–144 h postoperation. At the same time, clusterin mRNA levels rose severalfold. Thus, the increase in clusterin expression coincided with the occurrence of cell death (Rennie et al. 1988). Studying regression of human mammary cancer following estrogen withdrawal, Kyprianou and coworkers inoculated MCF-7 human mammary adenocarcinoma cells into ovariectomized female nude mice supplemented with exogenous estrogen to produce growing tumors (Kyprianou et al. 1991). They showed that within 1 day after hormone ablation, the MCF-7 cells display the morphological and biochemical changes indicative for apoptosis. At the same time they detected a threefold increase in clusterin and $TGF\beta1$ mRNA levels. While $TGF\beta1$ mRNA levels persisted through day 3 and then decreased, clusterin mRNA levels increased fivefold by 3 days' post-hormonal ablation and then slightly decreased. Both studies, however, did not analyze which cells account for the rise in clusterin gene expression, the cells undergoing apoptosis or those destined to survive.

A particularly well-characterized system is the regulation of clusterin gene expression in the prostate after castration. Castration-induced androgen deprivation causes a decrease in the nuclear androgen receptor content in the prostatic cells to undetectable levels as early as 12 h after treatment. DNA fragmentation and Ca^{2+}, Mg^{2+}-dependent DNase activity are observed very early, and increase during the first 24 h. At this time, apoptotic bodies accumulate and reach a maximum 2–3 days after castration (Buttyan 1991). The apoptotic bodies are eliminated by the phagocytic action of neighboring cells, local macrophages, and macrophages recruited from outside the prostate (Helminen and Ericson 1972; English et al. 1989). The analysis of mRNAs induced by castration revealed that the expression of a number of enzymes such as tissue and urokinase-type plasminogen activators, tissue transglutaminase, as well as poly(ADP-ribose) polymerase, the Yb1 subunit of gluthathione S-transferase, as well as clusterin (TRPM-2) is up-regulated in response to castration (Montpetit et al. 1986; Bettuzzi et al. 1989; Sensibar et al. 1991; Tenniswood et al. 1994). Of these, the upregulation of clusterin (TRPM-2) has been most extensively studied. It has been shown that clusterin expression in the male repoductive tract is not restricted to the areas of apoptosis and, therefore, is not unique to the process. In the involuting prostate tissue, the luminal epithelial cells in the central and distal regions of the ducts die, while those in the proximal region survive. During this process, the expression of clusterin is restricted to the luminal epithelial cells in the central and distal regions, where it is seen prior to and during recognizable morphological and biochemical changes associated with apoptosis (Buttyan et al. 1989; Buttyan 1991; Tenniswood et al. 1994). Therefore, it appears that clusterin gene expression is induced only in the cells located right in the center of ongoing apoptosis. On the basis of these observations it has been suggested that in the regressing prostate, clusterin gene expression is induced in the cells destined to die (Tenniswood, pers. comm). Because of its expression in normal Sertoli cells, however, it is unlikely that clusterin is causally involved in prostate apoptosis. If in the regressing prostate, clusterin is in fact expressed by the dying cells, one could rather imagine that this may represent a futile attempt of the dying cells to escape cell death or represent a cellular response to protect vital neighbor cells by inhibiting nonspecific complement activation and/or ensuring removal of cell debris.

4 Clusterin Gene Expression in the Thymus

In the lymphoid system, apoptosis is a frequent and physiological process. In the thymus, apoptosis has been demonstrated to be the mechanism used to extinguish potentially autoreactive thymocytes during negative selection (Smith et al. 1989; Dent et al. 1990; MacDonald and Lees 1990). Since clusterin gene expression had been shown to be enhanced in epithelial tissue regressing as a consequence of apoptosis, these findings initiated a number of investigations studying the expression of the clusterin gene in the thymus.

In one set of experiments, Bettuzzi and coworkers analyzed the expression of the clusterin gene in rat thymocytes during dexamethasone-induced apoptosis (Bettuzzi et al. 1991). Following a single intraperitoneal dexamethasone injection into adult male rats, they observed an increase in clusterin mRNA in isolated thymocytes as early as 30 min after glycocorticoid injection, which reached its maximum at 4 h and ceased at about 8 h. On the basis of their results, Bettuzzi and colleagues proposed that the protein might play an important role in thymocyte apoptosis.

A different conclusion was drawn by French and coworkers from their investigation on the expression of clusterin in four human thymuses collected from children under 2 years of age undergoing cardiac surgery (French et al. 1992).

Using in situ hybridization and immunocytochemistry, they showed that clusterin gene expression was confined to the epithelial cells present in the thymic medulla, with the cells surrounding the Hassal's bodies expressing the highest level. They did not detect clusterin expression in thymocytes. Their result strongly argues against a causal role of clusterin gene expression in the apoptotic processes occurring during T-cell selection. The authors rather propose that clusterin secreted by the medullary epithelial cells may help to dispose of cell debris generated by thymocyte apoptosis and to inhibit complement action in order to minimize the cytolytic damage of neighboring cells.

The same group extended their studies using in vitro models of apoptosis including dexamethasone induced apoptosis in a pre-T-lymphocyte cell line (French et al. 1994b). They could show that in none of their model systems did the morphological and biochemical changes characteristic for ongoing apoptosis coincide with an increase in clusterin gene expression. On the contrary, they observed a marked decrease or absence of clusterin gene expression. In accordance with their observations in human thymus, they demonstrated that clusterin gene expression in the in vitro systems was confined to morphologically normal cells.

These results were also supported by a study of clusterin gene expression in the rat thymus after in vivo dexamethasone administration (French et al. 1994a). As in normal human thymus, clusterin expression was exclusively detected in the medullary epithelial cells in the thymuses of dexamethasone-treated rats. The distribution of clusterin-expressing cells and the level of expression was unchanged during an 8-h interval after glucocorticoid administration, indicating that thymocyte apoptosis does not change the rate or site of clusterin gene expression in the thymus.

In conclusion, these data demonstrate that in the apoptotic thymus clusterin expression is confined to the nonapoptotic epithelial cells. This virtually rules out a causative role of the protein in the apoptotic process. Rather it appears to exert a protective function on the vital bystander cells. This protective role might be due to the protein's inhibitory action on the complement system, to its membrane binding activity and/or its ability to modulate cell-cell interactions.

5   Clusterin Gene Expression in Retinitis Pigmentosa (RP) Retina

Expression of clusterin has been implicated in inherited retinal degenerations including retinitis pigmentosa (RP), a major cause of human blindness. RP results in extensive degeneration of the photoreceptor cells, disruption of retinal architecture, accumulation of pigment-laden cells and proliferation of glial elements within the retina (Jomary et al. 1993 a,b). In a number of cases, mutations of genes specifically expressed in photoreceptor cells have been identified as the causal genetic defects in this disease (Humphries et al. 1992).

The comparison of the transcriptional activity in normal and RP-affected human retinas demonstrated an increase in the amounts of the mRNAs coding for glial fibrillary acidic protein (GFAP) and for clusterin (Jones et al. 1990, 1992). In the normal retina, clusterin mRNA was localized to the inner nuclear and ganglion cell layers. The strongest immunoreactivity for the protein was detected at the inner limiting membrane, plexiform layers, and outer segments of photoreceptors. Clusterin associated with the outer segments was presumably secreted by the overlying pigment epithelium since no clusterin mRNA was detected in the photoreceptors. By contrast, in RP-affected retinas, the organized distribution of clusterin is lost. Clusterin mRNA was found to be distributed throughout the RP-affected retina, whereas overall immunoreactivity was lower, indicating increased turnover, removal, or uptake and degradation of the protein within cells.

Recent studies in rds mutant mice support these conclusions. In rds mutant mice, photoreceptor cell differentiate normally only during the very first post-natal days and then undergo a slow degeneration, leading to the loss of the entire photoreceptor population presumably due to apoptotic cell death. In the dys-trophic retina of the mutant mice, clusterin mRNA was found to be overexpressed three-to four fold. As in normal human retinas, clusterin mRNA in the retinas of control mice and the rds mice was localized to the ganglion cells, the inner nuclear layers and in the retinal pigment epithelium. Unlike the human RP retinas, in the rds mutant mice, immunoreactive cells were seen scattered in the outer nuclear layer where the cell bodies of the photoreceptors reside, suggesting an association of clusterin with the dying photoreceptor cells, an observation that would be in accord with a function of clusterin in removing the debris of dying cells (Agarwal, pers. comm.).

The expression of clusterin in normal retinal tissue argues that the protein is not a causative agent in photoreceptor apoptosis. In the normal retina its distribution – predominantly in the synaptic layers, the outer limiting membrane, and the outer segments of the photolreceptors – correlates well with sites of high membrane turnover. In the RP-affected retina, clusterin may associate with the dying cell membrane to enhance the disposal of cytotoxic cell debris, and thereby to protect surviving neighboring cells.

6   Clusterin Expression During Degenerative Processes in the Brain

As in other tissues, apoptosis plays a role in the developing brain and can be induced in cultured neurons by the removal of growth factors, such as nerve growth factor (NGF) (Bredesen 1994). In the adult nervous system, too, apop-tosis occurs under certain conditions such as a sudden withdrawal of glucocorticoids following adrenalectomy and steroid replacement which leads to widespread cell death in the denate granule cells of the hippocampus (Sloviter et al. 1993). In addition, the widespread damage following ischemia or excito-toxicity can include apoptopic cell death of mature neurons (Kure et al. 1991). The involvement of apoptosis in neurodegenerative diseases has been difficult to determine because of the slow rate of neuronal cell death under these conditions;

however, recent work suggests that at least some neurodegenerative processes, including Alzheimer's disease, display biochemical hallmarks of apoptosis (Duguid et al. 1989).

Clusterin was identified by a number of groups by virtue of its elevated level of expression during neurodegenerative processes. The spectrum of pathological conditions in the brain associated with enhanced levels of clusterin includes Morbus Alzheimer's (May et al. 1990), Pick's disease (Duguid et al. 1989; Yasuhara et al. 1994), retroviral infection (Michel et al. 1989), prion diseases (Duguid et al. 1989), malign transformation (Danik et al. 1991), as well as numerous experimental lesions (Table 2).

One of the first findings linking clusterin expression to neurodegeneration came from Duguid and coworkers, who reported increased levels of the clusterin transcript in the RNA isolated from the brain of three Alzheimer's disease patients and one Pick's disease hippocampal sample (Duguid et al. 1989). In a parallel study, May and colleagues also found clusterin to be overexpressed in the brain of an Alzheimer's disease patient (May et al. 1990). The mRNA transcript was twofold elevated in comparison to normal controls. In addition, clusterin was localized to neurofibrillary tangle-bearing neurons and amyloid plaques. The origins of the clusterin-containing deposits remains unclear, but one possibility is that they are derived from local astrocytes and neurons not undergoing apoptosis. Taking into account the association of clusterin with lipids and membranes, the increased expression in Alzheimer's disease may reflect an involvement of clusterin in providing membrane precursors during neuronal remodelling. Furthermore, the localization of components of the complement system to amyloid plaques, tangle-bearing neurons, and dystrophic neurits may be indicative of a stimulated immune response in the Alzheimer's disease brain. Thus, the enhanced production of clusterin (complement lysis inhibitor) in the Alzheimer's disease brain could constitute a reaction of the surviving cells to control complement-mediated inflammatory processes in the affected area. It is not known, however, whether the protein found in the CNS displays an inhibitory action on complement-mediated cell lysis.

Table 2. Expression of clusterin in neurodegenerative processes

Insult	Site	Expression
M. Alzheimer	CA1 pyramidal and hilar neurons	Elevated (2-fold)
Epilepsy	Epileptic foci	Elevated
Glioma	Human glioma	Elevated (11-fold)
RSV infection	Quail neuroretinal cells	Elevated
Scrapie	Hamster brain	Elevated (10-fold)
BSE	Bovine brain	Elevated
Intraventricular kainate injection	CA3 and CA4	Elevated (2–4-fold)
Entorhinal cortex lesion	Sprouting in molecular layer of dentate gyrus	Elevated (2–4-fold)
Four-vessel occlusion	CA1	Elevated (2–4-fold)

Analyzing overexpressed genes in a human malignant astrocytoma, Danik and colleagues (1991) isolated a clone (pTB16) which, upon sequence analysis, was revealed to be the human homologue of clusterin. They found, through Northern analyses and in situ hybridization, that the level of expression of clusterin mRNA correlated with specific stages and types of brain tumors, and also noted high levels of expression in human epileptic foci.

Experimental lesions of the brain – whether surgical, chemical, or ischemic-also result in increases in clusterin mRNA (Table 2). Of particular interest are the studies of various hippocampal lesions that induce the death of selected neurons, such as application of the excitotoxin kainic acid, ischemia produced by four-vessel occlusion, and prepubertal adrenalectomy (May and Finch 1992).

All these results point to a possible association between enhanced levels of clusterin expression and neuronal cell death. Recent experiments are beginning to unravel the role of clusterin in these processes by answering such questions as whether clusterin is synthesized by dying neurons, surviving neurons, or glia; and how closely correlated is its spatial and temporal pattern of expression with sites of neuronal death.

Following up their previous results linking clusterin expression with brain tumors, Danik et al. examined the expression of clusterin mRNA in the normal adult rat brain and looked at the effect of the excitotoxin kaininc acid (Danik et al. 1993). They found that clusterin mRNA is expressed throughout the normal brain with a strong hybridization signal appearing in several and associated structures including the spinal chord motor neurons, the lining of the ventricles, and various hypothalamic and brainstem nuclei. Clusterin-positive cells included neurons, astrocytes, and epithelial cells of the choroid plexus. Interestingly, two areas of relatively low constitutive clusterin expression, the striatum and the CA1 region of the hippocampus, were found to produce strong hybridization signals several days after administration of kainic acid. Upon closer observation, they found that overexpression of the clusterin transcript coincided with glial fibrillary acidic protein (GFAP) expression, indicating that the cells producing the high levels of clusterin were astrocytes. The finding that clusterin is expressed constitutively in healthy neurons demonstrates that there is not an obligatory link between clusterin expression and neuronal cell death. Furthermore, the observation that the source of enhanced clusterin expression in response to excitotoxic stimuli is the astrocytes rather than the dying neurons eliminates the possibility of apoptopic up-regulation of clusterin expression in these neurons.

Similar results were obtained by Pasinetti et al. (1994), who investigated clusterin expression in the brain at both the mRNA and protein levels. Although clusterin mRNA was readily dectectable in almost all areas of the brain by in situ hybridization, most neurons and astrocytes in intact brain were immunonegative for the protein. This may be explained by the fact that constitutive secretion of clusterin could deplete intracellular stores to undetectable levels. In addition, they showed, using monotypic primary cultures, that astrocytes and neurons, but not microglia, express clusterin. Again, no convincing correlation was found in the neuronal distribution of clusterin and its possible role in cell death and apoptosis.

Using in situ hybridization, Wiessner et al. (1993) compared the time course of enhanced clusterin mRNA expression following ischemia to that of neuronal cell death. Ischemia was produced by four-vessel occlusion lasting 30 min followed by arterial recirculation for times ranging from 15 min to 7 days. Neuron loss in the dorsolateral striatum was observed within 12 h, and in the CAl region of the hippocampus after 2 to 3 days. Induction of clusterin mRNA was observed after 12 h; however, it was confined to reactive astrocytes. At no time point examined was there an induction of clusterin mRNA in neurons destined to die. Since clusterin is a secreted protein, the possibility remains, however, of an interaction with the plasma membrane of the dying cells.

Neuronal apoptosis can be observed during the development of the nervous system. Therefore, analysis of the pattern of clusterin expression during development should demonstrate conclusively whether or not there is a link between clusterin expression and apoptosis. Garden and colleagues have studied the expression of clusterin during normal development of the CNS and its correlation with the incidence of apoptosis. They compared, by in situ hybridization, the levels of clusterin mRNA in the developing and adult rat brain, and showed that clusterin mRNA is expressed in motor, cortical, and hypothalamic neurons at developmental stages when apoptosis has already terminated (Garden et al. 1991). Likewise, the distribution of clusterin in the murine-developing nervous system was examined by O'Bryan's group by immunohistochemistry and mRNA analysis (O'Bryan et al. 1993). They found that clusterin expression was first detected in the earliest neurons of the cortical plate on embryonic day 12. Thereafter, clusterin expression continued to rise with age, reaching a peak level in the post-natal mature brain. Virtually all neurons were found to be immunopositive for clusterin, and there was no specific induction of clusterin expression during epochs of apoptosis in the embryo. Rather than having a role in cell death, these studies suggest that clusterin may function in the maturation of neurons.

The recent data plainly rule out an obligatory link between clusterin expression and apoptosis in the CNS. Nonetheless, its prevalent expression in the brain implies that it does have an important function. The enhanced clusterin expression in degenerative RP retinas, in the CNS under senescence-related stresses, as well as in neurodegenerative processes triggered by exogenous insults such as infection, ischemia, and surgical trauma is likely to be a general response common to all these stresses. What might this function (or functions) be? The above results suggest that, rather than being involved in cell death, clusterin plays a role in the survival and maintenance of neurons. In keeping with this hypothesis, clusterin expression was found to be induced in the hippocampus by transforming growth factor $\beta1$ (TGF$\beta1$), a protective agent against ischemia in the brain, as well as important factor in inflammation and wound healing (Laping et al. 1994). Interestingly, TGF$\beta1$ immunoreactivity is also found in plaques of Alzheimer's disease brains (Van der Wal et al. 1993). Laping et al. have proposed that TGF$\beta1$ exerts its protective function, at least in part, through induction of clusterin expression.

At the biochemical level, the finding that it participates in hydrophobic interactions with lipids (Jenne et al. 1991) has led to the proposal that clusterin is involved in plasma membrane organization as a lipid transporter. This interpretation would be consistent with its increased levels of expression in maturing neurons in which extensive remodeling of the plasma membrane occurs during synaptogenesis. The low level of clusterin expression in the normal adult brain may be required for ongoing, background levels of synaptic turnover. Its overexpression in astrocytes in response to ischemia or treatment with kainic acid could aid in removal of debris from degenerating neurons by recycling membrane components. Thus, rather than being a precursor to cell death, clusterin appears to be required for the maintenance of healthy neruons.

7   Conclusions and Perspectives

The initial observation that clusterin was upregulated in the regressing prostate after castration raised several important questions:

– Is an enhancement of clusterin expression specific for the prostate or is it generally observed in regressing/apoptotic tissues?
– Is (high-level) clusterin expression strictly linked to apoptosis in a way that it could serve as an indicative (diagnostic) marker?
– If so, is clusterin expression causally involved in apoptosis?
– What are the molecular functions that clusterin exerts during apoptosis/ regression?

Within the last few years, answers, or at least clues, have been given to all of these questions. We know that up-regulation of clusterin is not unique to the regressing prostate, but occurs in the thymus, in a number of neurodegenerative processes, in liver and kidney injury, and carcinogenesis. It has also become clear that clusterin expression is not specific to apoptotic cells. On the contrary, it appears that in most tissues with ongoing apoptosis clusterin expression is restricted to the surviving bystander cells. These results clearly demonstrate that the function of clusterin in apoptosis is not causal; rather a protective role for the vital neighboring cells has been suggested. This prediction can be experimentally tested. The availability of highly purified clusterin as well as specific antibodies allows the determination of dose-effect relationships in vitro of treatments known to elicit apoptosis. Likewise the availability of cell lines stably expressing clusterin from a recombinant gene (Pilarsky et al. 1993; Appel and Koch-Brandt 1994) allows the comparative analysis of clusterin expressing and nonexpressing cell clones in order to test if the protein can exert a protective function on the producer cells.

The studies of clusterin function may be greatly enhanced by the recent identification of a cell surface clusterin receptor (Kounnas et al. 1995). This receptor was isolated on the basis of its homology to the low density lipoprotein (LDL) receptor. It binds clusterin with high affinity in a pH- and Ca^{2+}-dependent manner. The expression of the receptor is restricted to adsorptive epithelia, where it is localized in the luminal plasmamembrane domain. The receptor is also present in the chorioid plexus. Its level is elevated 100-fold in F9 embryonal carcinoma cells upon retinoic acid/cAMP-induced differentiation. These findings suggest that the function of clusterin at the luminal surfaces of lining epithelial cells, in the CNS, and during differentiation and morphogenesis may be mediated by its cell surface receptor. In apoptosis, the association of clusterin with the dying cell may promote the phagocytic uptake and degradation of cell debris by vital neighboring cells expressing the receptor.

Whatever clusterin's (prime) function may be, it has become clear that, although associated with apoptosis, enhanced clusterin expression is not strictly coupled to the process. Therefore, it cannot be regarded as a reliable marker for apoptosis. Rather, it may be indicative for more general membrane turnover and remodeling processes.

Acknowledgments. We thank our colleagues and collaborators, especially Dr. W. Bursch, for contributing results prior to publication and for stimulating discussion and valuable advice. We thank further Carmen Scholz and Christina Weindel for help with the preparation of the manuscript.

References

Appeal D, Koch-Brandt C (1994) Sorting of a secretory protein (gp80) to the apical surface of Caco-2 cells. J Cell Sci 107: 553–559

Bettuzzi S, Hiipakka RA, Gilna P, Liao S (1989) Identification of an androgen-repressed mRNA in rat ventral prostate as coding for sulfated glycoprotein 2 by cDNA cloning and sequence analysis. Biochem J 257: 293–299

Bettuzzi S, Troiano L, Davalli P, Tropea F, Ingletti MC, Grassilli E, Monti D, Cori A, Franceschi C (1991) In vivo accumulation of sulfated glycoprotein 2 mRNA in rat thymocytes upon dexamethasone-induced cell death. Biochem Biophys Res Commun 175: 810–815

Bissonnette RP, Shi Y, Mahboubi A, Glynn JM, Green DR (1994) c-*myc* and apoptosis. In: Tomei LD, Cope FO (eds) Apoptosis II: the molecular basis of apoptosis in disease. Cold Spring Harbor Lab Press, New York, pp 327–356

Blaschuk O, Burdzy K, Fritz IB (1993) Purification and characterization of a cell-aggregating factor (clusterin), the major glycoprotein in ram rete testis fluid. J Biol Chem 258: 7714–7720

Brandyk M, Buttyan R, Olsson CA, Appeal G, D'Agati V, Katz A, Ng PY, Sawczuk IS (1990) TRPM-2 detection and localization during gentamicin-induced nephrotoxicity. J Urol 143: 239A

Bredesen DE (1994) Neuronal apoptosis: genetic and biochemical modulation. In: Tomei LD, Cope FO (eds) Apoptosis II: the molecular basis of apoptosis in disease. Cold Spring Harbor Lab Press, New York, pp 397– 421

Brown DG, Sun XM, Cohen GM (1993) Dexamethasone-induced apoptosis involves cleavage of DNA to large fragments prior to internucleosomal fragmentation. J Biol Chem 268: 3027–3039

Bursch W, Lauer B, Timmermann-Trosiener I, Barthel G, Schuppler J, Schulte-Hermann R (1984) Controlled cell death (apoptosis) of nonmal and putative preneoplastic cells in rat liver following withdrawal of tumor promoters. Carcinogenesis 5: 453–458

Bursch W, Oberhammer F, Schulte-Hermann R (1992a) Cell death by apoptosis and its protective role against disease. Trends Pharmacol Sci 13: 245–251

Bursch W, Fesus L, Schulte-Hermann R (1992b) Apoptosis ('programmed' cell death) and its relevance in liver injury and carcinogenesis. In: Dekant w, Neumann HG (eds) Tissue-specific toxicity: biochemical mechanisms. Academic Press, London, pp 95–115

Bursch W, Gleeson T, Kleine L, Tenniswood M (1995) Expression of clusterin/testosterone-repressed prostate message (TRPM-2) mRNA during growth and regression of rat liver. Arch Toxicol 69: 253–258

Buttyan R (1991) Genetic reponse of prostate cells to androgen deprivation: insights into the cellular mechanism of apoptosis. In: Tomei LD, Cope FO (eds) Apoptosis: the molecular basis of cell death. Cold Spring Harbor Lab Press, New York, pp 157–173

Buttyan R, Olsson CA, Pintar J, Chang C, Bandyk M, Ng PY, Sawczuk IS (1989) Induction of the TRPM-2 gene in cells undergoing programmed cell death. Mol Cell Biol 9: 3473–3481

Cheng YC, Mathur PP, Grima J (1988) Structural analysis of clusterin and its subunits in ram rete testis fluid. Biochemistry 27: 4079–4088

Choi-Miura NH, Takahashi Y, Nakano Y, Tobe T, Tomito M (1992) Identification of the disulfide bonds in human plasma protein SP-40, 40 (Apolipoprotein-J). J Biochem 112: 557–561

Collard MW, Griswold MD (1987) Biosynthesis and molecular cloning of sulfated glycoprotein 2 secreted by rat sertoli cells. Biochemistry 26: 3297–3303

Collins MKL, Rivas AL (1993) The control of apoptosis in mammalian cells. Trends Biochem 18: 307–309

Danik M, Chabot J-G, Mercier C, Benabid A, Chauvin C, Quirion R, Suh M (1991) Human gliomas and epileptic foci express high levels of a mRNA related to rat testicular sulfated glycoprotein 2, a purported marker of cell death. Proc Natl Acad Sci USA 88: 8577–8581

Danik M, Chabot J-G, Hassan-Gonzales D, Suh M, Quirion R (1993) Localization of sulfated glycoprotein-2/clusterin mRNA in the rat brain by in situ hybridization. J Comp Neurol 331: 209–227

Davidson NE, Lippman ME (1988) Treatment of metastatic breast cancer. In: Lippman ME, Lichter AS, Danforth DN (eds) Diagnosis and management of breast cancer. WB Saunders, London, pp 375–406

Dent AL, Matis LA, Hooshmand, F, Widacki SM, Bluestone JA, Herdick SM (1990) Self-reactive gamma delta T cells are eliminated in the thymus. Nature 343: 714–719

De Silva HV, Harmony JAK, Stuart WD, Gil CM, Robbins J (1990) Apolipoprotein J: structure and tissue distribution. Biochemistry 29: 5380–5389

Diemer V, Hoyle M, Baglioni C, Millis AJT (1992) Expression of porcine complement cytolysis inhibitor mRNA in cultured aortic smooth muscle cells. J Biol Chem 267: 5257–5264

Duguid JR, Bohmont CW, Liu N, Tourtellotte WW (1989) Changes in brain gene expression shared by scrapie and Alzheimer disease. Proc Natl Acad Sci USA 86: 7260–7264

Ellis RE, Yuan J, Horvitz HR (1991) Mechanisms and functions of cell death. Annu Rev Cell Biol 7: 663–698

English HF, Kyprianou N, Isaacs JT (1989) Relationship between DNA fragmentation and apoptosis in the programmed cell death in the rat prostate following castration. Prostate 7: 41–49

Filipski J, Leblanc, J, Youdale T, Sikorska M, Walker PR (1990) Periodicity of DNA folding in higher order chromatin structures. EMBO J 9: 1319–1327

French LE, Chonn A, Ducrest D, Baumann B, Belin D, Wohlwend A, Kiss JZ, Sappino A-P, Tschopp J, Schifferli JA (1993) Molecular cloning and mRNA localization of a gene associated with epithelial differentiation processes during embryogenesis. J Cell Biol 122: 1119–1130

French LE, Sappino A-P, Tschopp J, Schifferli JA (1992) Distinct sites of production of the putative cell death marker clusterin in the human thymus. J Clin Invest 90: 1919–1925

French LE, Sappino, A-P, Tschopp J, Schifferli JA (1994a) Clusterin gene expression in the rat thymus is not modulated by dexamethasone treatment. Immunology 82: 328–331

French LE, Wohlwend A, Sappino A-P, Tschopp J, Schifferli JA (1994b) Human clusterin gene expression is confined to surviving cells during in vitro programmed cell death. J Clin Invest 93: 877–884

Fritz IB, Burdzy K (1989) Novel action of carnitine: inhibition of aggregation of dispersed cells elicited by clusterin in vitro. J Cell Physiol 140: 18–28

Fritz IB, Burdzy K, Setchell B, Blaschuk O (1983) Ram rete testis fluid contains a protein (clusterin) which influences cell-cell interactions in vitro. Biol Reprod 28: 1173–1188

Furuya Y, Berges R, Lundmo P, Isaacs JT (1994) Cell proliferation, p53 gene expression, and intracellular calcium in programmed cell death: prostate model. In: Tomei LD, Cope FO (eds) Apoptosis II: the molecular basis of apoptosis in disease. Cold Spring Harbor Lab Press, New York, pp 231–252

Garden GA, Bothwell M, Rubel EW (1991) Lack of correspondence between mRNA expression for a putative cell death molecule (SGP-2) and neuronal cell death in the central nervous system. J Neurobiol 22: 590–604

Gougeon ML, Montagnier L (1994) Apoptosis in peripheral lymphocytes during HIV infection: influence of superantigens and correlation with AIDS pathogenesis. In: Tomei LD, Cope FO (eds) Apoptosis II: the molecular basis of apoptosis in disease. Cold Spring Harbor Lab Press, New York, pp 5–20

Griswold MD, Roberts K, Bishop P (1986) Purification and characterization of sulfated glycoprotein secreted by Sertoli cells. Biochemistry 25: 7265–7270

Hartmann K, Rauch J, Urban J, Parcyk K, Diel P, Pilarsky C, Appel D, Haase W, Mann K, Weller A, Koch-Brandt C (1991) Molecular cloning of gp80, a glycoprotein complex secreted by kidney cells in vivo. J Biol Chem 266: 9924–9931

Helminen HJ, Ericson JLE (1972) Ultrastructural studies on prostatic involution in the rat. Evidence for focal irreversible damage to epithelium and heterophagic digestion in macrophages. J Ultrastruct Res 39: 443–452

Hovanessian AG (1994) Apoptosis in HIV-infection: the rule of extracellular and transmembrane glycoproteins. In: Tomei LD, Cope FO (eds) Apoptosis II: the molecular basis of apoptosis in disease. Cold Spring Harbor Lab Press, New York, pp 21–42

Humphries P, Kenna P, Farrar GJ (1992) On the molecular genetics of retinitis pigmentosa. Science 256: 804–813

Huttner WB, Gerdes H-H, Rosa P (1991) The granin (chromogranin/secretogranin) family. Trends Biochem Sci 16: 27–30

Jenne DE, Tschopp J (1989) Molecular structure and functional characterization of a human complement cytolysis inhibitor found in blood and seminal plasma: identity to sulfated glycoprotein 2, a constituent of rat testis fluid. Proc Natl Acad Sci USA 86: 7123–7127

Jenne DE, Tschopp J (1992) Clusterin: the intriguing guises of a widely expressed glycoprotein. Trends Biochem Sci 17: 154–159

Jenne DE, Lowin B, Peitsch MC, Böttcher A, Schitz G, Tschopp J (1991) Clusterin (complement lysis inhibitor) forms a high density lipoprotein complex with apolipoprotein A-1 in human plasma. J Biol Chem 266: 11030–11036

Jomary C, Murphy BF, Neal MJ, Jones SE (1993a) Abnormal distribution of retinal clusterin in retinitis pigmentosa. Mol Brain Res 20: 274–278

Jomary C, Neal MJ, Jones SE (1993b) Comparison of clusterin gene expression in normal and dystrophic human retinas. Mol Brain Res 20: 279–284

Jones SE, Wood-Gush HG, Cunningham JR, Szczesny PJ, Neal MJ (1990) Abnormalities in expression of human retinal mRNA in retinitis pigmentosa. Neurochem Int 17: 495–503

Jones SE, Meerabux JMA, Yeats DA, Neal MJ (1992) Analysis of differentially expressed genes in retinitis pigmentosa retinas: altered expression of clusterin mRNA. FEBS Lett 300: 279–282

Jordan-Starck TC, Witte DP, Aronow BJ, Harmony JAK (1992) Apolipoprotein J: a membrane policeman. Curr Opin Lipidol 3: 75–85

Kadomatsu K, Anzano MA, Slayter V, Winokur TS, Smith JM, Sporn MB (1993) Expression of sulfated glycoprotein 2 is associated with carcinogenesis induced by N-nitroso-N-methylurea in rat prostate and seminal vesicle. Cancer Res 53: 1480–1483

Kerr JFR, Harmon BV (1991) Definition and incidence of apoptosis: an historical perspective. In: Tomei LD, Cope FO (eds) Apoptosis: the molecular basis of cell death. Cold Spring Harbor Lab Press, New York, pp 5–30

Kirszbaum L, Sharpe JA, Murphy B, d'Apice JF, Classon B, Hudson P, Walker ID (1989) Molecular cloning and characterization of the noval, human complement-associated protein, SP 40, 40: a link between the complement and reproductive systems. EMBO J 8: 711–718

Kirszbaum L, Bozas WE, Walker ID (1992) SP-40, 40, a protein involved in the control of the complement pathway, possesses a unique array of disulphide bridges. FEBS Lett 297: 70–76

Kounnas MZ, Loukinova EB, Stefansson S, Harmony JAK, Brewer BH, Strickland DK, Argraves WS (1995) Identification of glycoprotein 330 as an endocytic receptor for Apolipoprotein J/Clusterin. J Biol Chem 270: 13070–13075

Kure S, Tominaga T, Yoshimoto T, Tada K, Narisawa K (1991) Glutamate triggers internucleo-somal DNA cleavage in neuronal cells. Biochem Biophys Res Commun 179: 39–46

Kyprianou N, English HF, Isaacs JT (1988) Activation of a $Ca^{++-Mg^{++}}$-dependent endonuclease as an early event in castration-induced prostate cell death. Prostate 13: 103–109

Kyprianou N, English HF, Isaacs JT (1990) Programmed cell death during the regression of PC-82 human prostate cancer following androgen ablation. Cancer Res 50: 3748–3753

Kyprianou N, English HF, Davidson NE, Isaacs JT (1991) Programmed cell death during regression of the MCF-7 human breast cancer following estrogen ablation. Cancer Res 51: 162–166

Laping NJ, Morgan TE, Nichols NR, Rozovsky I, Young-Chan Cs, Zarow C, Finch CE (1994) Transforming growth factor-β1 induces neuronal and astrocyte genes: tubulin α1, glial fibrillary acidic protein and clusterin. Neuroscience 58: 563–572

Laslop A, Steiner H-J, Egger C, Wolkersdorfer M, Kapelari S, Hogue-Angeletti R, Erickson JD, Fischer-Colbrie R, Winkler H (1993) Glycoprotein III (clusterin, sulfated glycoprotein 2) in endocrine, nervous, and other tissues: immunochemical characterization, subcellular localization, and regulation of bisynthesis. J Neurochem 61: 1498–1505

Lee S, Christakos S, Small MB (1993) Apoptosis and signal transduction: clues to a molecular mechanism. Curr Op in Cell Biol 5: 286–291

MacDonald HR, Lees RK (1990) Programmed death of autoreactive thymocytes. Nature 343: 642–644

Martin SJ (1993) Apoptosis: suicide, execution or murder. Trends Cell Biol 3: 141–144

May PC, Finch CE (1992) Sulfated glycoprotein 2: new relationships of this multifunctional protein to neurodegeneration. Trends Neurosci 15: 391–396

May PC, Lampert-Etchells M, Johnson SA, Poirier J, Masters JN, Finch CE (1990) Dynamics of gene expression for a hippocampal glycoprotein elevated in Alzheimer's disease and in response to experimental lesions in rat. Neuron 5: 831–839

May PC, Robinson P, Fuson K, Smalstig B, Stephenson D, Clemens JA (1992) Sulfated glycoprotein-2 expression increases in rodent brain after transient global ischemia. Mol Brain Res 15: 33–39

Michel D, Gillet G, Volovitch M, Pessac B, Calothy G, Brun G (1989) Expression of a noval gene encoding a 51.5 kD precursor protein is induced by different retroviral oncogenes in quail neuroretinal cells. Oncogene Res 4: 127–136

Montpetit ML, Lawless KR, Tenniswood M (1986) Androgen-repressed messages in the rat ventral prostate. Prostate 8: 25–36

Murphy BF, Kirszbaum L, Walker ID, d'Apice AJF (1988) SP-40, 40 a newly identified normal human serum protein found in the SC5b-9 complex of complement and in the immune deposits in glomerulonephritis. J Clin Invest 81: 1858–1864

Nagata S (1994) Apoptosis-mediating Fas antigen and its natural mutation. In: Tomei LD, Cope FO (eds) Apoptosis II: the molecular basis of apoptosis in disease. Cold Spring Harbor Lab Press, New York, pp 313–326

Oberhammer F, Wilson JW, Dive C, Morris ID, Hickman JA, Wakeling AE, Walker PR, Sikorska M (1993) Apoptotic death in epithelial cells: cleavage of DNA to 300 and/or 50 kb fragments prior to or in the absence of internucleosomal fragmentation. EMBO J 12: 3679–3684

O'Bryan MK, Cheema SS, Barlett PF, Murphy BF, Pearse MJ (1993) Clusterin levels increase during neuronal development. J Neurobiol 25: 421–432

Osborne BA, Schwartz LM (1994) Essential genes that regulate apoptosis. Trends Cell Biol 4: 394–399

Palmar DJ, Christie DL (1990) The primary structure of glycoprotein III from bovine adrenal meduallary chromaffin granules. J Biol Chem 265: 6617–6623

Parczyk K, Pilarsky C, Rachel U, Koch-Brandt C (1994) Gp80 (clusterin; TRPM-2) mRNA level is enhanced in human renal clear cell carcinomas. J Cancer Res Clin Oncol 120: 186–188

Pasinetti GM, Johnson SA, Oda T, Rozovsky I, Finch CE (1994) Clusterin (SGP-2): a multi-functional glycoprotein with regional expression in astrocytes and neurons of the adult rat brain. J Comp Neurol 339: 387–400

Peitsch MC, Mannherz HG, Tschopp J (1994) The apoptosis endonucleases: cleaning up after cell death? Trends Cell Biol 4: 37–41

Pignataro OP, Feng Z-M, Chen C (1992) Cyclic adenosine 3', 5' -monophosphate negatively regulates clusterin gene expression in Leydig tumor cell lines. Endocrinology 130: 2745–2750

Pilarsky C, Haase W, Koch-Brandt C (1993) Stable expression of gp80 (TRPM-2, clusterin), a secretory protein implicated in programmed cell death, in transfected BHK-21 cells. Biochim Biophys Acta 1179: 306–310

Purello M, Betuzzi S, DiPietro C, Mirabile E, Di Blasi M, Rimini R, Grzeschik KH, Ingletti C, Corti A, Sichel G (1991) The gene for SP-40, 40, human homolog of rat sulfated glycoprotein 2 rat clusterin, and rat testosterone-repressed prostatic message 2, maps to chromosome 8. Genomics 10: 151–156

Raff MC (1992) Social controls on cell survival and cell death. Nature 356: 397–400

Rennie PS, Bruchovsky N, Buttyan R, Benson M, Cheng H (1988) Gene expression during the early phases of regression of the androgen-dependent Shionogi mouse mammary carcinoma. Cancer Res 48: 6309 – 6312

Sawczuk IS, Hoke Olsson CA, Connor J, Buttyan R (1989) Gene expression in response to acute unilateral ureteral obstruction. Kidney Int 35: 1315–1321

Sensibar JA, Griswold MD, Sylvester SR, Buttyan R, Bardin CW, Cheng CY, Dudek S, Lee C (1991) Prostatic ductal system in rats: regional variation in localization of an androgen-repressed gene product, sulfated glycoprotein-2. Endocrinology 128: 2091–2102

Sloviter RS, Sollas AL, Dean E, Neubart S (1993) Adrenalectomy-induced granule cell degeneration in the rat hippocampal dentate gyrus: characterization of an in vivo model of controlled neuronal death. J Comp Neurol 330: 324–336

Smith CA, Williams GT, Kingston R, Jenkinson EJ, Owen JJT (1989) Antibodies to CD3/T cell receptor complex induce death by apoptosis in immature T cells in thymic cultures. Nature 337: 181–184

Smith CA, Grimes EA, McCarthy NJ, Williams GT (1994) Multiple gene regulation of apoptosis: Significance in immunology and oncology. In: Tomei LD, Cope FO (eds) Apoptosis II: the molecular basis of apoptosis in disease. Cold Spring Harbor Press, New York, pp 43–88

Tenniswood M, Taillefer D, Lakins J, Guenette R, Mooibroek M, Daehlin L, Welsh J (1994) Control of gene expression during apoptosis in hormone-dependent tissues. In: Tomei LD, Cope FO (eds) Apoptosis II: the molecular basis of apoptosis in disease. Cold Spring Harbor Lab Press, New York, pp 283–311

Tschopp J, Jenne DE, Hertig S, Preissner KT, Morgenstern H, Sapino A-P, French L (1993) Human megakaryocytes express clusterin and package it without apolipoprotein A-1 into α-granules. Blood 82: 118–125

Urban J, Parczyk K, Leutz A, Kayne M, Kondor-Koch C (1987) Constitutive apical secretion of an 80-kD sulfated glycoprotein complex in the polarized epithelial Madin-Darby canine kidney cell line. J Cell Biol 105: 2735–2743

Van der Wal EA, Gomez-Pinilla F, Cotman CW (1993) Transforming growth factor-β1 is in plaques in Alzheimer and Down pathologies. Neuro Rep 4: 69–72

Wadewitz AG, Lockshin RA (1988) Programmed cell death: dying cells synthesize a co-ordinated, unique set of proteins in two different episodes of cell death. FEBS Lett 241: 19–23

White E, Gooding LR (1994) Regulation of apoptosis by human adenoviruses. In: Tomei LD, Cope FO (eds) Apoptosis II: the molecular basis of apoptosis in disease. Cold Spring Harbor Lab Press, New York, pp 111–142

White K, Grether ME, Abrams JM, Young L, Farrell K, Steller H (1994) Genetic control of programmed cell death in Drosophila. Science 264: 677–683

Wiessner C, Back T, Bonnekoh P, Kohno K, Gehrmann J, Hossmann KA (1993) Sulfated glyco-protein-2 mRNA in the rat brain following transient forebrain ischemia. Mol Brain Res 20: 345–352

Williams GT (1991) Programmed cell death: apoptosis and oncogenesis. Cell 65: 1097–1098

Williams GT, Smith CA (1993) Molecular regulation of apoptosis: genetic controls on cell death. Cell 74: 777–779

Williams GT, Smith CA, McCarthy NJ, Grimes EA (1992) Apoptosis: final control point in cell biology. Trends Cell Biol 2: 263–267

Witte DP, Aronow BJ, Staudermann ML, Stuart WD, Clay MA, Gruppo R, Jenkins SH, Harmony JAK (1993) Platelet activation releases megakaryocyte-synthesized apolipoprotein J, a highly abundant protein in atheromatous lesions. Am J Pathol 143: 763–773

Wong P, Pineault JM, Lakins J, Taillefer D, Leger JG, Wang C, Tenniswood MP (1993) Genomic organisation and expression of the rat TRPM-2 (clusterin) gene, a gene implicated in apoptosis. J Biol Chem 268: 5021–5031

Wyllie AH (1980) Glucocorticoid-induced thymocyte apoptosis is associated with endogenous endonuclease activation. Nature 284: 555–556

Wyllie AH (1985) The biology of cell death in tumors. Anticancer Res 5: 131–136

Yasuhara O, Aimi Y, McGeer E, McGeer P (1994) Expression of the complement membrane attack complex and its inhibitors in pig-disease brain. Brain Res 652: 346–349

Springer-Verlag
and the Environment

We at Springer-Verlag firmly believe that an international science publisher has a special obligation to the environment, and our corporate policies consistently reflect this conviction.

We also expect our business partners – paper mills, printers, packaging manufacturers, etc. – to commit themselves to using environmentally friendly materials and production processes.

The paper in this book is made from low- or no-chlorine pulp and is acid free, in conformance with international standards for paper permanency.

Printing: Saladruck, Berlin
Binding: Buchbinderei Lüderitz & Bauer, Berlin